MODELLING NITROGEN FROM FARM WASTES

Models and Systems for Studying the Transformation and Fate of Nitrogen from Animal Effluents Applied to Soils

Proceedings of a Seminar in the EEC Programme of Coordination of Research on Animal Effluents, organised by Professor H. Laudelout, and held at the Université Catholique de Louvain, Louvain-la-Neuve, Belgium, October 10–11, 1978.

The Seminar was sponsored by the Commission of the European Communities, Directorate-General for Agriculture, Coordination of Agricultural Research.

MODELLING NITROGEN FROM FARM WASTES

Models and Systems for Studying
the Transformation and Fate of Nitrogen
from Animal Effluents Applied to Soils

Edited by

J. K. R. GASSER

Agricultural Research Council, London, UK

APPLIED SCIENCE PUBLISHERS LTD
LONDON

APPLIED SCIENCE PUBLISHERS LTD
RIPPLE ROAD, BARKING, ESSEX, ENGLAND

British Library Cataloguing in Publication Data
Modelling nitrogen from farm wastes.
1. Liquid fertilizers and manures—
Mathematical models   2. Farm manure—
Mathematical models   3. Soils—Nitrogen content
—Mathematical models
I. Gasser, John Kenneth Russell
631.4′2   S662

ISBN 0-85334-869-3

WITH 27 TABLES AND 66 ILLUSTRATIONS

© ECSC, EEC, EAEC, Brussels and Luxembourg 1979

Publication arrangements by: Commission of the European Communities, Directorate-General for Scientific and Technical Information and Information Management, Luxembourg

EUR 6361 EN

LEGAL NOTICE
Neither the Commission of the European Communities nor any person acting on behalf of the Commission is responsible for the use which might be made of the following information.

All rights reserved. No part of this publication may be reproduced, stored in a retrieval system, or transmitted in any form or by any means, electronic, mechanical, photocopying, recording, or otherwise, without the prior written permission of the publishers, Applied Science Publishers Ltd, Ripple Road, Barking, Essex, England

Printed in Great Britain by Galliard (Printers) Ltd, Great Yarmouth

CONTENTS

Preface vii

FIRST SESSION

THE USE OF OPEN SOIL SYSTEMS AND MATHEMATICAL
MODELS TO STUDY N REACTIONS AND MOVEMENT IN SOILS
J.L. Starr, J-Y. Parlange and D.R. Nielsen.. .. 1

CONVERSION AND DEGRADATION OF ORGANIC AND
INORGANIC NITROGEN COMPOUNDS FROM HEAVILY
LOADED WASTE WATERS IN AMPHIBIC SOILS
R. Kickuth 35

DISCUSSION 41

THE EFFECT OF NITROGEN IN PIG SLURRY SPREAD
OUT AT DIFFERENT APPLICATION TIMES
H. Vetter and G. Steffens 44

MATHEMATICAL MODELLING OF COMPLEX REACTIONS
AND SOLUTE MOVEMENT THROUGH SOILS
J. Dufey 62

DISCUSSION 71

SECOND SESSION

THE USE OF TRACERS TO DETERMINE THE DYNAMIC
NATURE OF ORGANIC MATTER
E.A. Paul and J.A. van Veen 75

MATHEMATICAL MODELLING OF NITROGEN
TRANSFORMATIONS IN SOIL
J.A. van Veen and M.J. Frissel 131

NITRATE PRODUCTION, MOVEMENT AND LOSSES
IN SOILS
J.K.R. Gasser 158

DISCUSSION 169

NITROGEN MINERALISATION AND NITRIFICATION OF
PIG SLURRY ADDED TO SOIL IN LABORATORY CONDITIONS
J.C. Germon, J.J. Giraud, R. Chaussod
and C. Duthion 170

THIRD SESSION

DISCUSSION 187

CONCLUSIONS AND RECOMMENDATIONS 190

LIST OF PARTICIPANTS 193

PREFACE

The Expert Group on Animal Effluents of the Standing Committee on Agricultural Research organised a number of full seminars during the 1976 - 1978 programme dealing with several aspects of livestock effluents. The subjects discussed included analytical methods, odour problems, health questions and engineering requirements for effluent treatment and disposal. Many people either state explicitly or assume tacitly that much of the effluent from livestock will eventually be returned to the land. The effects of untreated and treated effluent on crop growth and the potential for pollution depend on a large number of factors. Their importance, quantification and interactions may be best described by the use of mathematical and biological models. The present seminar brought together experts in the various aspects of modelling and soil processes to assess the present position and needs for future work. The organisers hope these aims have been fulfilled.

FIRST SESSION

Chairman: R. Lecomte

THE USE OF OPEN SOIL SYSTEMS AND MATHEMATICAL MODELS TO STUDY N REACTIONS AND MOVEMENT IN SOILS

J.L. Starr[1], J.-Y. Parlange[2] and D.R. Nielsen[3]

[1] Department of Soil and Water,
The Connecticut Agricultural Experimental Station, New Haven, USA.

[2] School of Australian Environmental Studies, Griffith University, Brisbane, Australia.

[3] Department of Land, Air and Water Resources, University of California, Davis, USA.

The fate of nitrogen in soils, especially in relation to plant growth, has been studied extensively and intensively throughout the world for nearly a century. Public interest in the fate of N in soils has expanded from agricultural interests to include those of general environmental importance. For example, an entire issue of the new international journal, AMBIO (Arrhenius, 1977), was devoted to the nitrogen cycle, and in the USA, the National Science Foundation and the Environmental Protection Agency have greatly increased their funding of N research projects. Much of the past N research has been in the form of fertiliser trials, with an emphasis on maximising crop productivity (Ensminger and Pearson, 1950; Allison, 1966). Laboratory studies have often treated the soil as a closed system to elucidate the factors controlling N reactions and transformations. The field studies have resulted in the development of basic guidelines for agronomic practices and the laboratory studies have provided important qualitative information regarding N in soils. However, with the rising costs of N fertilisers and the increasing concern for the environment, qualitative information alone is insufficient. Research must be directed more toward a better understanding of the kinetics of microbial and chemical transformation of inorganic- and organic-N compounds in the soil and their attendant micro-environmental parameters amenable to manipulation and control.

Reactions and transformations of nitrogen are controlled by many physical, chemical and biological properties of the soil; and occur simultaneously or sequentially in both space and time (Macura and Kunc, 1965; McLaren, 1970; Misra et al., 1974; Starr et al., 1974). These studies of open systems which link the chemical and physical properties of soils with the growth and metabolic processes of soil micro-organisms are fundamental to the N cycle. Such experiments, coupled with appropriate mathematical models, can provide the necessary information and tools to examine and explain the rates at which various processes within the N cycle take place, to identify control measures for improving the quality of surface and groundwaters and to increase the uptake of fertiliser N by plants.

Many mathematical models have been developed to describe the displacement of a fluid and its dissolved constituents through a porous medium. A general equation that has proved to be particularly useful in laboratory studies to define and analyse the one-dimensional leaching characteristics of soils in the presence of reversible ion exchange reactions and irreversible microbiological transformations is

$$\frac{\partial s_i}{\partial t} + \frac{\partial c_i}{\partial t} = D_i \frac{\partial^2 c_i}{\partial x^2} - v_i \frac{\partial c_i}{\partial x} + \emptyset_i \qquad (1)$$

where c_i is the concentration of solute i in the soil solution (ML^{-3}), not identified with the adsorbed phase; s_i is the concentration of solute i in the soil solution identified with the adsorbed phase; D_i is the apparent diffusion coefficient of the solute (L^2T^{-1}); v_i is the average interstitial velocity of the solution (LT^{-1}); and $\emptyset_i$ is the net time rate at which the mass of solute i is being irreversibly produced or used per unit volume of solution due to a specific transformation. Most solutions to Equation 1 assume that the values of D_i and v_i are invariant while x and t are the space and time co-ordinates, respectively. Note that if i represents a cation (such as ammonium) entering into reversible ion exchange reactions, $\partial s_i/\partial t \neq 0$. If however, i represents an anion which does not appreciably react with soil particle surfaces, $\partial s_i/\partial t = 0$. The term $\emptyset_i$ represents a net irreversible rate of transformation occurring in the solution phase alone. Examples of $\emptyset_i$ include absorption of the solute by plant roots, microbiological transformations of the solute stemming from additions of fertilisers or other additions to the soil including crop and solid waste residues, and, in the case of a radioactive solute, its decay. Analytical and numerical solutions of Equation 1 are available for a large number of initial and boundary conditions (Boask, 1973; Kirda et al., 1973; Misra et al., 1974; Starr and Parlange, 1975; Wagenet et al., 1977). These solutions will not be reviewed, the purpose is rather to show how this basic model has been used in the design and analysis of my own research on the reaction and transport of N and other dissolved constituents in and through soils.

LABORATORY EXPERIMENTS

This first example illustrates how measurements of nitrogen as well as other elements essential for microbial growth in the liquid and gaseous phases of a soil being leached with a solution of an ammonium salt can be analysed quantitatively from chemical, physical and microbiological points of view simultaneously.

A sandy-loam surface-soil was sieved and packed into an acrylic plastic column 15.2 cm x 13.2 cm x 100 cm. The uniformity of packing and the distribution of soil water throughout the column was measured by gamma radiation attenuation. Small fritted-glass tubular-cups and porous-nylon tubing were placed in the column at 5 cm intervals for extracting the soil solution and the soil air respectively. Soil-water pressure at the top and bottom of the column was regulated to maintain an air content of 10% during the entire leaching process. The soil column, maintained at 20 ± 0.5°C, was steadily leached with 0.01 $\underline{N}$ CaSO$_4$, to maintain the soil colloids in a flocculated state, with an average pore-water velocity of 2.5 mm/h. The tracer treatment consisted of suddenly (at a time designated as zero) starting to enrich the leaching solution with 50 mg/l

NH_4^+-N, labelled with 8.65 atom percent excess ^{15}N. Samples of soil solution and soil air were taken periodically and analysed for NH_4^+, NO_2^- and NO_3^- and N_2, N_2O, O_2, and CO_2, respectively. This treatment was continued for 39 days, until the measured ionic and gaseous concentrations in the soil were constant with time. Owing to the acidifying effect of nitrification, dolomitic limestone was then applied to the soil surface at an equivalent rate of 4.5 tonnes/ha and leached for 56 days with 0.01 $\underline{N}$ $CaSO_4$. The first treatment (Experiment A) was then repeated (Experiment B), in an attempt to reproduce the conditions and results obtained; with a new steady state developing 41 days. Additional experimental details are available (Starr et al., 1974).

Figure 1 shows measured and theoretical distributions of NH_4^+-N and NO_3^--N within the soil column after 39 days and 41 days for the two experiments, A and B. Nitrite was never present in detectable amounts (less than 0.1 mg/l). It is apparent that NH_4^+ is rapidly oxidised in the upper 30 cm of the soil profile, and that the NO_3^- being formed is subsequently reduced even near the soil surface. The rate of reduction is sufficiently great to keep the NO_3^- concentration below 35 mg/l, or 30% less than the 50 mg/l expected had no denitrification occurred.

To evaluate the transformation term $\emptyset_i$ in Equation 1, v and D must first be measured in the absence of reaction (eg, $\emptyset_i(x) = 0$). For these steady flow experiments, v and D were determined by sending a pulse of Cl^- into the column, with v measured directly and D obtained by fitting the observed depth and time distributions with the appropriate solution to Equation 1(Kirkham and Powers, 1972; Nielsen and Biggar, 1962; Warrick et al., 1971). Experiments have previously shown that movement of Cl^- approximates that of NO_3^- (Corey et al., 1967). The value of D for NH_4^+ was assumed to be the same as for nitrate for the steady state conditions (eg $\partial s_i/\partial t = 0$ in Equation 1).

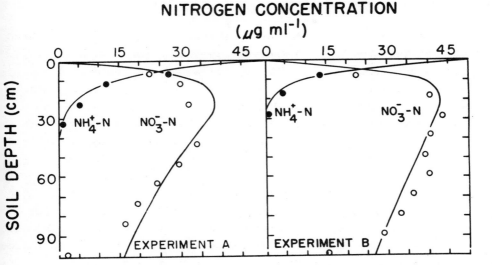

Fig. 1. Steady state NH_4^+-N and NO_3^--N concentration distributions with soil depth for Experiment A and Experiment B. The smooth curves were drawn with the solution to Equation 1.

The concentrations of the nitrogenous constituents in the soil solution were analysed using Equation 1 rewritten for steady state conditions and the consecutive reactions,

$$NH_4^+ \xrightarrow{k_1} NO_3^- \xrightarrow{k_2} N_2, \text{ or}$$

$$D \frac{\partial^2 c_1}{\partial x^2} - v \frac{\partial c_1}{\partial x} = k_1 c_1 \qquad (2)$$

and

$$D \frac{\partial^2 c_2}{\partial x^2} - v \frac{\partial c_2}{\partial x} + k_1 c_1 = k_2 c_2 \qquad (3)$$

where c_1 and c_2 are the respective concentrations of NH_4^+ and NO_3^-, and the terms forming $k_1 c_1$ are $\emptyset_1$ where it is assumed that (i) the number of microbes oxidising NH_4^+ is constant and (ii) oxidation of NH_4^+ occurs as a first order reaction. Similarly, the number of soil microbes denitrifying NO_3^- is constant and reduction of NO_3^- occurs as a first order reaction.

The slope of the line drawn through a plot of the left hand side of Equation 2 for measured values of D and v with c_1 at each data point gives k_1 (Figure 2); k_2 may be found in a similar manner. These values were then used in the analytical solution to Equation 1 to draw the theoretical lines through the results in Figure 1. The excellent fit of the theory with the experimental results suggests that our assumption of first order kinetics is indeed correct. However, since Figure 2 shows that the actual kinetics appear to be first order over only a small range of concentrations, we must be a little cautious about our belief that these kinetics are first order. Steady profiles have since been shown to give reliable information only about some integral of the kinetics rather than the detailed kinetics even when the microbial distribution is measured independently (Starr and Parlange, 1976a). In this particular case an acceptably good fit can also be obtained assuming zero order kinetics. Closed system studies indicate that the K_M for nitrification is about 5 mg/l, so that, at the concentration used in this experiment the reaction should be zero order (McLaren, 1975). Again some caution should be exercised on the use of this value of K_M since closed system studies also indicate that nitrification is negligible at values more acid than pH 5.0 (Dancer et al., 1973). However, near the soil surface at the end of experiment B the acidity was as great as pH 4.37, and the rate of nitrification was certainly not negligible. Hence, the measurement of K_M's and other microbial parameters from closed systems cannot be extended automatically to open system experiments.

Equation 1 can also assist us in the analysis of the measurement of soil gases in the soil atmosphere. For steady state conditions and the value of v being zero (the movement takes place only by gaseous diffusion), Equation 1 reduces to

$$\emptyset_i(x) = -D_i \frac{\partial^2 c_i}{\partial x^2} \qquad (4)$$

with known values of the apparent gaseous diffusion coefficient D_i and the concentration distributions of the gases N_2, CO_2 and O_2, $\emptyset_i(x)$, the rate at which each gas is consumed or produced, can be found as a function of soil depth. This approach is illustrated in Figure 3B from the labelled N_2 distributions shown in Figure 3A. The values of the gaseous diffusion coefficient D for N_2 was based upon measurements of the concentration gradient and flux of CO_2 at the soil surface (eg from Fick's law, $q_i = D_i (\partial c/\partial x)$). The $^{15}N_2$ is produced by the reduction of nitrate formed as a result of the oxidation of the added NH_4^+. Similarly, by measuring the O_2-distribution the rate of O_2 consumption from the soil air by the nitrifying organisms was compared stoichiometrically with the rate of disappearance of NH_4^+ from the soil solution as a function of soil depth. CO_2 behaviour was treated in the same manner. In this experiment, the use of isotopically labelled N plus the analysis of both the solution and gaseous phases of the soil with time and depth provided the means for identifying and quantifying sources and sinks in terms of stoichiometric and active biomass relations.

A second example illustrates the research method for a much smaller column to investigate the more narrowly defined objective of how quickly and at what rate an inherent soil microbial population will reduce NO_3^--N following a rapid decline in the O_2 status of the soil, without previously assuming either a first- or zero-order kinetic model (Starr and Parlange, 1975). To achieve this objective it was necessary to measure quickly the rate relative to microbial growth kinetics, since any continuing experiment will alter the number and composition of the soil microbes. The experiment reported here was completed within 1 day. (The method is now being extended to studies of microbial kinetics for much shorter time spans.) It was also necessary to solve Equation 1 without assuming the functional relationship between the reaction rate and the nitrate concentration.

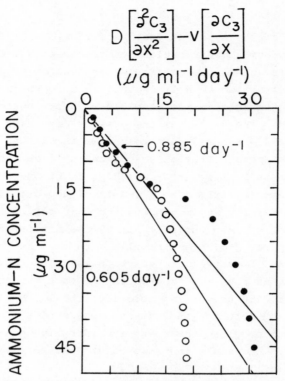

Fig. 2. The left hand side of Equation 2 and the concentration of NH_4^+-N in the soil solution. Open and solid circles designate measured values for experiments A and B, respectively. The straight lines represent the average slopes, k_1.

An undisturbed core of a sandy-loam surface soil was taken by pushing an acrylic perforated column, with an inside diameter of 6 cm and a length of 10 cm, into the soil. The column was removed and porous plates were placed on each end of the column through which the leaching solution and leachate were passed. The column of field moist soil was placed in a concentric outer cylinder with the air space between the cylinder blocked off on opposite sides (Figure 4). This allows for continuous passage of a water-saturated gas, with or without O_2, through the unsaturated soil perpendicular to the flow of the soil solution. Hence, the soil atmosphere could be made aerobic or anaerobic at will.

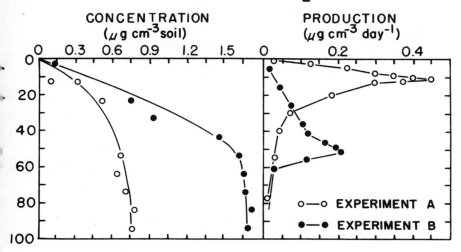

Fig. 3. Labelled N_2 concentration (Fig. A) and production rate stemming from Equation 4 (Fig. B) and soil depth.

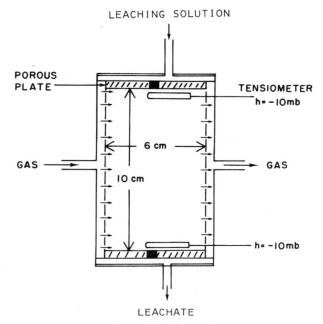

Fig. 4. Schematic diagram of column for continuous vertical flow of solution and horizontal flow of atmosphere.

A constant flow of 0.01 $\underline{N}$ $CaSO_4$ in tap water was established with the soil-water pressure maintained at -10 mbars at the top and bottom of the column. The initial treatment consisted of adding 50 ppm NO_3^--N under aerobic conditions until the concentration of NO_3^- in the leachate and leaching solution were the same. The initial breakthrough curve for NO_3^--N under aerobic conditions is shown in Figure 5. The average pore water velocity, v, was measured directly and the convective diffusion D, was determined by fitting Equation 1 to the data, with $\emptyset_i = 0$

Soluble carbon was previously found to be readily leached from the column during this phase of the experiment. Hence, in the present experiment, after achieving a NO_3^--N concentration of 50 ppm throughout the column (in the absence of added carbon) the gas passing through the column was abruptly switched to N_2 and the leaching solution was enriched with glucose-carbon at a concentration of 35 ppm. This is enough carbon to denitrify about three-quarters of the added NO_3^--N if the denitrifiers are the only consumers of carbon (Broadbent and Clark, 1965).

The resultant relative NO_3^--N concentration with time in hours (or pore volumes) is shown in Figure 6. Note that the NO_3^--N concentration drops precipitously following T = 8 hours, or one pore volume. Owing to the provison of carbon at the top of the column, any denitrification that might occur could not be observed for nearly 8 hours (one pore volume) with the pore-water velocities of this experiment. The fact that it dropped precipitously at that time indicates that the soil denitrifiers began using NO_3^--N in respiration with an insignificant delay following the switch to anaerobic conditions. The value of the reaction term $\emptyset_i(c)$ was then computed by curve fitting our solution to Equation 1 using the previously measured values of v and D.

It should be emphasised that in this study we were interested in establishing a method without a prior assumption of either a first-order or zero-order kinetic model. As an

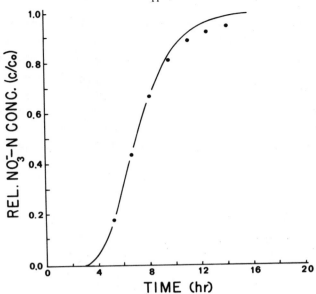

Fig. 5. Measurement of D. The solution to Equation 1 was fitted to the breakthrough curve in the absence of denitrification, $\phi_i = 0$, to measure D_i.

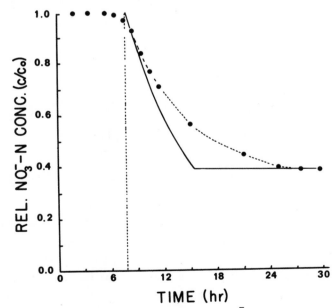

Fig. 6. Breakthrough curve obtained by sending NO_3^- and sugar through a column containing a solution of NO_3^- but no sugar and switching to anaerobic conditions at the same time.

illustration of the procedure, the rate of denitrification, $\emptyset_i(c)$ with change of NO_3^--N concentration, is shown in Figure 7. Indeed the reaction kinetics are first order at concentrations less than 0.8 C/C_o, or 40 ppm NO_3^--N, but at greater concentration an apparent higher order reaction occurs. For $0.6 < C/C_o < 0.8$, $\emptyset_i$ is quite linear and extrapolates to $\emptyset(0) = 0$. As expected for low values of NO_3^- and/or soluble carbon the reaction is first order, and the extrapolation indicated in Figure 7 is reliable.

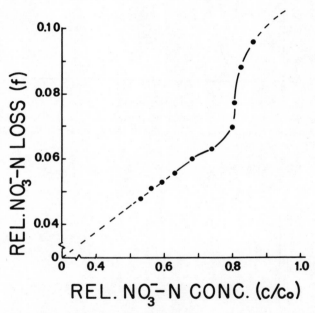

Fig. 7. Values of f obtained from the experiment reported in Figure 6. Extrapolated values are indicated by the dotted line for $C/C_o < 0.5$ and $C/C_o > 0.8$.

The kinetic nature of the denitrification reaction thus determined and quantified was used to ascertain how well these measured reaction rates can be used to predict the NO_3^--N distribution under a different set of conditions. Hence, after nearly all the previously applied NO_3^--N had been displaced with NO_3^--free solution, NO_3^--N at 35 ppm with the same carbon concentration but with a decreased pore water velocity was added to the soil surface under anaerobic conditions.

The results of that experiment (not shown) adequately described the new conditions and clearly demonstrated that the data clearly falls between the two extremes, and hence, can only be described as a mixed order regardless of the exact value of D.

Under natural or field conditions there may be times that it is safe to assume a first or zero order kinetic model. However, most often we will not know beforehand what the reaction kinetics are. Hence, the experimental method and associated theoretical model presented here represent a fast method for determining denitrification rates of an inherent soil microbial population without assuming a particular kinetic model.

This same experimental technique has also been used successfully by Volz and Starr (1977) to measure the denitrifier population dynamics with respect to time and depth and the specific rates of denitrification per bacterium. The next experiment to be done utilising this technique will also use ^{13}N (with a half-life of 10 minutes) as an extremely sensitive tracer on the movement of NO_2^- and gaseous anaerobic products (Smith et al., 1978).

Owing to natural heterogeneity, direct quantitative use of Equation 1 is difficult in field research, although some significant results have been achieved (Biggar and Nielsen, 1976; Nielsen et al., 1973). However, because this is a mechanistic model, it does serve to remind the experimenter that the reversible and irreversible reactions of interest are time and depth dependent and are directly related to the time, depth and spatial variations in the hydraulic properties of the soil. Hence it is an aid in the planning and design of field experiments.

FIELD EXPERIMENTS

The first field illustration is from an experiment designed to assess the fate of N and C in a septic system draining into the sub-soil in the field through a soakaway (Starr and Sawhney, 1979).

A 6-year old septic system serving a household of 4 adults was used in this study. The system was designed so that the effluent from the septic tank can be diverted to either of two parallel leaching trenches. Each trench consists of a series of 3 precast leaching chambers (1.22 m x 2.44 m x 0.3 m) placed end-to-end at a depth of 1.4 m in the coarse sand sub-soil and surrounded with 0.3 m of gravel (Figure 8). Large holes on all sides of the precast leaching chambers allow effluent to move out horizontally into the gravel and soil when ponding occurs.

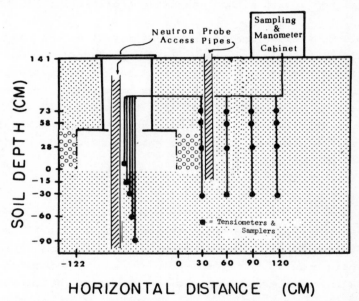

Fig. 8. Cross-sectional end view of the instrumented drain-field. The distances are given relative to the edge and bottom of the leaching chamber.

Since installation, each trench had been used alternately for 6 months. In March 1975, after one of the trenches had remained unused for about 6 months, a total of 40 porous ceramic probes (20 tensiometers and 20 soil water samplers) and 2 neutron probe access pipes were placed below and to the side of one of the leaching chambers.

The septic effluent was directed to the instrumented soakaway from May to October 1975, and again from May of 1976 until 1978. Soil water samples for N, C and P analysis and tensiometer and neutron probe readings were taken approximately weekly for 3 consecutive years. Partial results from the two replicate seasons, May to November of 1975 and 1976, are presented below.

Septic system effluent began to pond in the leaching chamber within 24 hours after the effluent was directed to the trench. The depth of ponded effluent increased until the hydraulic head, H, was great enough to maintain a rate of infiltration approaching that of the imposed effluent loading rate (i.e. $H \simeq 90$ cm in 100 days). Five months later the effluent was diverted to the second trench, and the first trench was allowed to dry until the following spring when the cycle was repeated.

Soil-water pressure data from below the leaching chamber (Figure 9), together with the soil moisture release curves for this soil (not shown), show that the soil below the leaching chamber was essentially saturated to a depth of 60 cm in 1975 and to a depth of 30 cm in 1976. This difference in soil water, pressure between 1975 and 1976 may be due to rainfall in 1975 being much greater than in 1976. (The normal monthly rainfall for the period June to December is 98 ± 10 mm, whereas the observed monthly rainfall for 1975 and 1976 was 150 ± 55 mm, and 105 ± 73 mm, respectively.) The observation that the soil below the ponded slime layer in 1976 was unsaturated is consistent with the findings of other experimenters (Bouma et al., 1972).

The volumetric water contents below the chamber, determined by the neutron probe (Figure 10) also showed considerable differences for the two years. In 1975 the volumetric water content, θ, showed little change for nearly two months after the effluent was diverted to the instrumented soakaway, and then began to fluctuate widely. The water content in the soil profile

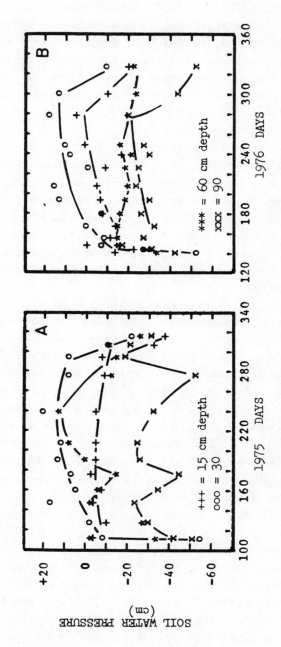

Fig. 9. Soil water pressure distributions with time at 15, 30, 60 and 90 cm depths below the leaching chamber in 1975 and 1976.

was maximal around day 190, becoming nearly saturated at the 60 cm depth. In 1976 the soil profile approached saturation immediately following ponding and then decreased to a water content of about 30% at depths of 30 cm and 60 cm, and about 15% at 90 cm below the trench.

The soil-water data from this experiment varied so greatly that the pattern of water movement in the leaching field is not readily apparent nor does it permit direct calculation of the mass of N moving through the soakaway. For example, the SWP data would indicate that the water potential gradient is both up and down at the 30 cm depth during much of the period of measurement. Clearly, the water is not moving through this soil in a simple one-dimensional pattern. Others have observed that as a saturated water front moves downwards from a layer of low hydraulic conductivity K_s into a layer with a much greater K_s, an instability of flow develops at the interface of the two layers and the flow breaks into narrow fingers, i.e. changing from one-dimensional to three-dimensional flow (Hill and Parlange, 1972; Starr et al., 1978; Takagi, 1960).

This non-uniformity of flow below the leaching chamber is also apparent from a comparison of the soil-water pressure and water-content data. In 1975, the soil-water content at the tensiometers was consistently at or near saturation at soil depths from 15 to 60 cm, whereas the soil near the neutron probes varied greatly over the same depths and time period. Only at the depth of 90 cm did the two methods give similar results. Conversely, in 1976 the soil-water content near the neutron probe varied less with time than it did near the tensiometers. Some of these differences are undoubtedly due to the difference in volume of soil measured by the two methods. Nonetheless, this comparison clearly shows that the soil-water content below the leaching chamber varies greatly both horizontally and vertically.

Concentrations of NH_4^+-N and NO_3^--N below the leaching chamber show considerable differences between 1975 and 1976

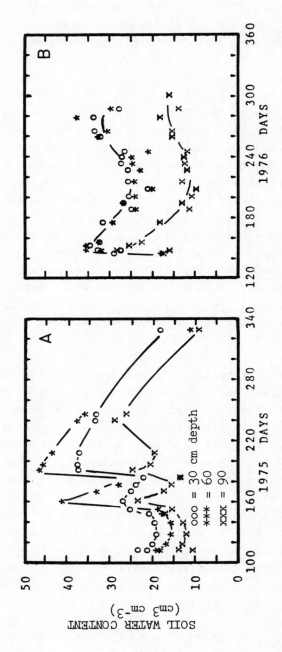

Fig. 10. Volumetric soil-water content distributions with time at 30, 60 and 90 cm depths below the leaching chamber.

(Figure 11), which may be due to the difference in rainfall for the two years as noted earlier.

NH_4^+-N in the effluent increased rapidly from 0 to 12 ppm in the first 15 days of 1975, and then slowly to about 25 ppm corresponding to a typical exponential bacterial growth curve with approximately 15 days being required for the N-mineralising bacteria in the effluent to reach half its maximal population. However, in 1976 the NH_4^+-N concentration quickly rose to its maximum suggesting that little if any time lag was required for bacterial growth as in the previous year. Greater concentrations of NH_4^+-N in the soil solution below the slime layer than in the ponded effluent indicate that O_2 was available for mineralisation in and below the slime layer as well as in the ponded effluent above.

In contrast to 1975, concentrations of NH_4^+-N in 1976 below the leaching chamber did not tend to constant values at all depths. Instead the concentration dropped abruptly to 1 - 3 ppm at depths of 60 and 90 cm and with a concomitant rise in NO_3^--N concentration (Figure 11D). The pulse of NO_3^--N at 60 cm and 90 cm deep about day 180 is the result of nitrification occurring in the profile from the end of 1975 to May of 1976 (Figure 11A and B), and its subsequent movement in the infiltrating water. After about 3 months of leaching, the NO_3^--N at depths of 60 cm and 90 cm tended to become 25 ppm.

Annual rainfall is clearly important in the fate of N in this septic system soakaway. Indeed, opposite conclusions with respect to NO_3^--N movement to the groundwater could be drawn if the data were limited to just one year. Ammonium-N tends to become constant at all depths in 1975 suggesting that O_2 was not available for nitrification in 1975, with the NH_4^+-N moving to depths below 90 cm at concentrations of approximately 25 ppm. On the other hand, more O_2 seems to be available throughout the soil profile in 1976. Much of the NO_3^--N observed in the NO_3^--N pulse (Figure 11D) was probably produced prior to diverting the flow back to the instrumented soakaway. However, Figure 12

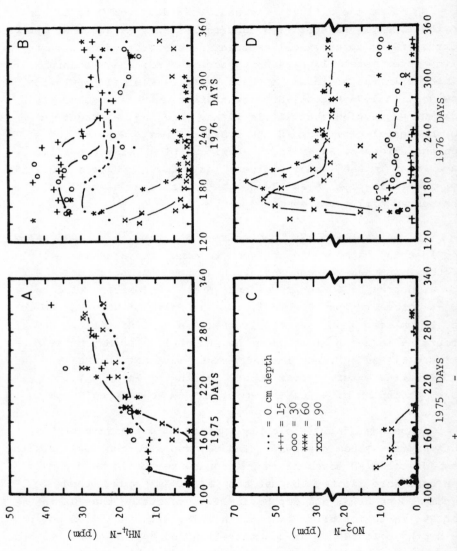

Fig. 11. Concentration of NH_4^+-N and NO_3^--N with time at 0, 15, 30, 60 and 90 cm depths below the leaching chamber

shows that the continued oxidation of organic materials below the slime layer is indicated by the high concentrations of soluble organic C at a depth of 30 cm.

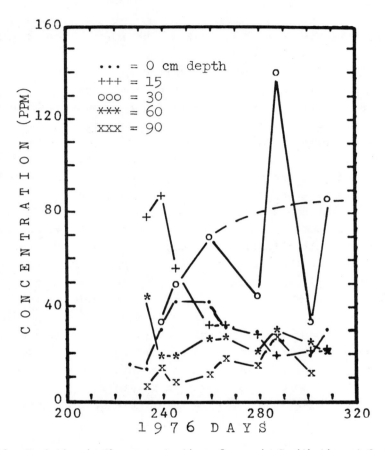

Fig. 12. Variation in the concentration of organic-C with time at 0, 15, 30, 60 and 90 cm below the leaching chamber.

Determinations of soluble organic-C in samples of solution were made starting in August 1976. Carbon concentrations in the ponded effluent fluctuated between 10 and 40 ppm and approached a constant value of 25 ppm. A similar pattern was observed for samples taken from 15, 60 and 90 cm below the chamber with the concentration in the samples from 15 cm deep fluctuating

around 60 ppm in the middle of the summer. Samples from the
30 cm depth, however, were unlike those above or below, showing
a much wider range in concentrations and fluctuating around a
concentration of 80 ppm.

Although it is evident that a true steady state does not
exist, the trends are apparent. A concentration profile of C,
NH_4^+-N and NO_3^--N after day 190 more clearly indicates the
general microbial activity at this time (Figure 13). The
NH_4^+-N and organic-C curves in this figure show rapid mineralisation of organic material at depths from 15 to 30 cm. The
concomitant rise in NO_3^--N between 15 and 60 cm indicates the
general aerobic nature of the soil in this region. With no
denitrification, the maximum steady state concentration of
NO_3^--N would be equal to the maximum NH_4^+-N. The NO_3^--N curve
indicates that the abrupt drop in organic-C below 30 cm must
be due to an active aerobic-heterotrophic bacterial population
and not to denitrification. Hence, it is clear that in 1976,
essentially all the mineralised-N below the chamber is nitrified
and then quantitatively transported to depths below 90 cm and
probably moves quantitatively to the water table. With
approximately equal concentrations of C and NO_3^--N at a depth of
90 cm it is possible that nearly four-fifths of the NO_3^--N could
be reduced in an anaerobic zone deeper in the soil. However,
the fact that the concentration of organic-C becomes constant
between 60 and 90 cm indicates that metabolism of C has
essentially ceased and suggests that there are few bacteria at
this depth. Hence, it is improbable that the number of
denitrifiers at depths greater than 90 cm would be sufficient
to lessen the mass of NO_3^--N delivered to the groundwater from
this system.

We conclude from these quasi steady state data that there
is sufficient O_2 in the septic system soakaway to mineralise
about one-quarter of the total effluent N, even in unusually
wet years. In years with average rainfall or less, this
NH_4^+-N will be nitrified and move to the groundwater. In this
system, it appears that much of the remaining N in the effluent

accumulates in the leaching reservoir until the system becomes aerobic due to diverting the effluent to the other leaching trench. At that time, the mass of organic-N may be largely mineralised and nitrified resulting in a massive pulse of nitrate-N leaching through the soil upon diverting the effluent back to this line.

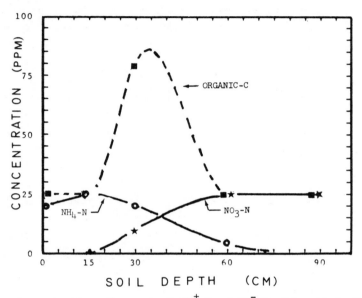

Fig. 13. Concentration of organic-C, NH_4^+-N and NO_3^--N with soil depth below the leaching chamber for days 280 - 320 in 1976.

Owing to the time and spatial variability in the soil water data, we had to use quasi steady state profiles to estimate the mass of N moving to the groundwater. The possibility for direct quantification would have been greatly enhanced by an experimental design that would have allowed for true three-dimensional measurement and analysis of the soil water. This together with periodic applications of tracers (e.g. Cl^-, Br^-, $^3H^+$) to the effluent, would have permitted a stoichiometric analysis of the soil water, N and C in the soakaway and a clearer picture of the time and depth dependent factors controlling the fate of N in this soakaway.

Other field experiments have been, and are being done using ^{15}N depleted or enriched nitrogenous fertilisers to study the fate of N in and below the crop rooting zone. We have previously seen that studies of the fate of N in soil columns are greatly facilitated by characterising the convective-dispersive properties of the soil in the absence of reactions. This presentation concludes with a brief review of a study illustrating a method for characterising hydraulic properties of the field soils on which these labelled N studies are being conducted (Starr et al., 1978).

The site for these leaching studies has a fine sandy-loam soil which overlies a coarse sand at a depth of 80 - 100 cm, with a water table at about 200 cm. Laboratory studies on layered soils suggested that water movement at the textural interface may become unstable, breaking into fingers of flow through the coarse sand below (Hill and Parlange, 1972). Hence, one of the objectives of this study was to determine if such an instability in the movement of water also occurs in this layered field soil and to observe its effect(s) upon the pattern of solute movement during leaching. This field study was in two parts; the first utilised a large in situ soil column and the second a suction lysimeter.

In the first part, a steel cylinder 1.8 m in diameter was driven into the soil to a depth of 3.6 m to provide a soil column for two ponded-flow infiltration experiments. The first experiment examined the rate of infiltration when the entrapped air could only escape through the saturated surface layer. The second experiment examined water and dye movement in the absence of entrapped air by infiltrating 45 cm of water containing a green vegetable dye. After infiltration of this dye solution successive layers of soil were removed from the cylinder and the dye pattern of each newly exposed surface was photographed.

In this latter experiment distinct dye patterns in the soil were clearly evident in the coarse subsoil. At 1 m deep there

were 12 distinct green zones (Figure 14) which were all
located on one side of the 1.8 m diameter column. These green
spots ranged from 5 to 20 cm in diameter and occupied only 5%
of the total cross-sectional area. Upon further removal of the
soil, each green spot was observed to continue with depth to
the water table. The free water at 2.1 m was seen to be dark
green beneath the area where the green fingers of soil were
most numerous but was clear where they were absent.

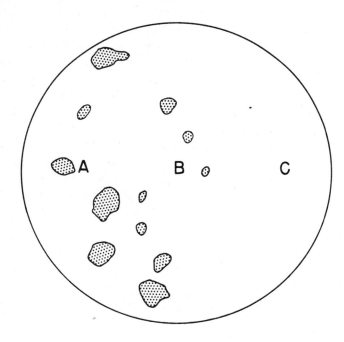

Fig. 14. Schematic cross-section of the 1.8 m soil column at the 100 cm
depth, drawn from a colour photograph. The dotted areas represent
the zones of dark green colour. Water samples at a depth of 210 cm
had optical densities as follows: Position A = 0.42, B = 0.05,
C = 0.00. The optical density of the dye solution applied was 0.88
(all at 550 μm).

This experiment demonstrates that ponded flow of water
through this layered, fine over coarse, soil produced an
instability with most of the water moving through the coarse
subsoil in small fingers of flow rather than as a one-dimensional
infiltration front.

The second part of this field experiment was designed to study solute movement through the same layered soil using field techniques described by Biggar and Nielsen (1976). Four plots, each 4.6 m x 6.1 m adjacent to the experiments in Part I, were instrumented with two neutron probe access pipes, two suction probes, and one tensiometer at depths of 20, 40, 60, 120, 180, 240 and 300 cm as shown in Figure 15. Details of the probe design and installation are available (Starr et al., 1978). After the probes were installed, the plots were levelled and ponded to a depth of 5 cm with water until a steady water content and pressure were observed, using the neutron probes and tensiometers, throughout the 300 cm profile. The 5 cm of water was then allowed to infiltrate the soil and a 5 cm pulse of 0.3 $\underline{N}$ $CaCl_2$ solution was applied to the soil surface. Upon infiltration of this solution, the plots were again ponded to and maintained at a depth of 5 cm with $CaCl_2$-free water.

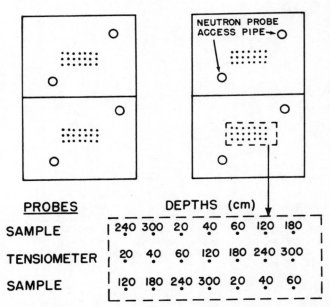

Fig. 15. Schematic location of the probes and access pipes in each of the four plots. Each 4.6 m by 6.1 m plot has two neutron probe access pipes, plus two suction probes and one tensiometer at each of the seven soil depths shown.

The four plots reached an apparent steady state with respect to the soil-water profile within 48 hours. The soil-water content profile exhibited a decreasing water content to a depth of 130 cm and then increased steadily to the saturated zone at 200 cm. The steady state flux, q of the infiltrating ponded water in the four plots were 27, 18, 18 and 16 mm/h.

The relative Cl^- concentration with depth and time following the application of the 5 cm pulse of 0.3 $\underline{N}$ $CaCl_2$ solution are shown in Figure 16. Note that several salt pulses reached depths of 120 cm and 180 cm very soon after they had reached 60 cm. The difference between the mean time of travel of the solute from 60 cm to 180 cm deep divided by the difference in depth produces a solute velocity, v_s, that is an order of magnitude greater than the calculated saturated pore water velocity, v_w, near the soil surface. This observed rapid movement of solute through the coarse layer is in complete agreement with the dye patterns observed in part I. Indeed, this comparatively high v_s corresponds to the high flux density required to transport all the infiltrating water through the observed dye fingers in Experiment I. Such a high flow rate is indeed possible owing to the relative coarseness of this layer. Hence, at depths greater than 60 cm, these experiments show that both the water and solute must be moving in discrete fingers and therefore more rapidly through this layer than would have occurred with simple one-dimensional flow.

Under these field conditions, the assumption of one-dimensional movement of water throughout the soil profile can lead to gross error in estimating the amount of dissolved constituents that are transported to groundwater. Owing to the extensiveness of natural conditions in which the conductivity of the subsoil is greater than the surface layer above it, rigorous laboratory and field studies are needed if we are to predict accurately the contributions of surface-applied materials to groundwater aquifers.

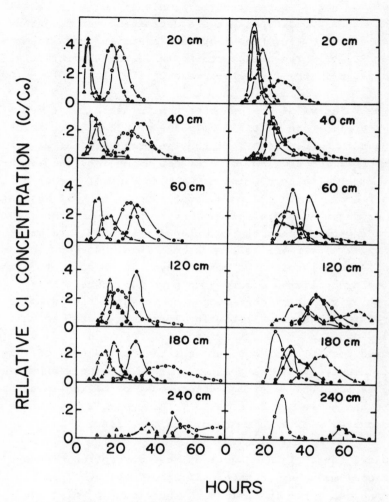

Fig. 16. Relative chloride concentration with time, measured at six soil depths for the eight suction probes in the four plots. Symbols of like shape represent the two samples from the same plot. The left part of the figure gives the concentration curves corresponding to plots with q = 2.7 (▲, △) and 1.8 (●, O) cm/h, respectively. Similarly, the right part corresponds to plots with q = 1.8 (▲, △) and 1.6 (●, O) cm/h.

We should keep in mind that tremendous variabilities in the movement of soil water and solutes are not unique to layered soils. Biggar and Nielsen (1976) have been studying the flux of water and solutes in soils with deep 'homogeneous' soil profiles and have observed that while measurements of static features of the soil (e.g. soil-water content) have a normal distribution, the dynamic features (e.g. pore-water velocities) vary with a log-normal distribution, as illustrated in Figure 17 which shows that the mean pore-water velocity cannot be used to characterise or describe the movement of water and solute through this soil unless the frequency distribution of the observations are also thoroughly analysed.

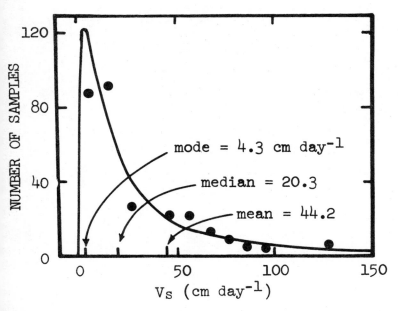

Fig. 17. Frequency distribution of values of the pore water velocity, v_s, for a class length of 10 cm/d.

The above laboratory and field studies illustrate the use of open soil systems and mechanistic models to observe and analyse the fate of N in soils in terms of various reaction rates integrated over space and time. The model presented here should not be considered the best or ultimate model, it is a first approximation of reality. There is much yet to be learned regarding the reaction and transport of N in soil with the mathematical models helping us to do more than just add more experimental data to our note books. With the challenges that lie ahead, especially regarding the spatial variability that occurs in nature, we must not withdraw into the secure world of our own specialties or disciplines. Instead, it should foster interdisciplinary research while being ready to abandon traditional approaches and equations in pursuit of the models and concepts that will better enable us to describe and control the fate of N in the field.

APPENDIX

Laboratory techniques

Soil column experiments can provide a good approximation of the physical, chemical and biological conditions that occur in nature. However, observations from these experiments can be quite misleading because of poor experimental design and difficulties with the experimental apparatus. The latter is perhaps the more serious because it is often overlooked. For example, breakthrough curves (BTC) often show some asymmetry or tailing. Although this tailing may be associated with many physical and chemical phenomena, part or all of this tailing may be inadvertently caused by the method of packing soil in the column, unstable flow (Starr and Parlange, 1976b), and by irregularities in the flow field at the entrance of the column (e.g. cracked or dirty entrance plates or influent flow rates that are much less than the hydraulic conductivity of the entrance plate (Starr and Parlange, 1977).

The packing of soil columns to minimise layering and particle separation is more of an art than a science and often requires repacking the same column several times before doing the experiment. Careful examination of the porous plates (particularly for short columns) for occlusions, cracks or other defects is also required. Further, the hydraulic conductivity of the entrance plate should not be greater than the imposed solution flux.

Owing to the reaction kinetics involved, it is often desirable to do column experiments on quite small quantities of soil. However, it is generally not advisable to use columns of less than 2 cm in diameter due to the wall effect, and theoretically, the column must be infinite in length. Practically, the mathematical models are adequate for columns with a Peclet number (Pe = Lv/D) $\geq$ 16. For smaller values (e.g. 4 > Pe > 16) different boundary conditions must be used to solve Equation 1 (Starr and Parlange, 1975). Experimentally, the range of v/D values will often be from 1 to 20 for homogeneously packed soil columns. The mathematical models have yet to be verified to describe adequately the observations for columns with L < 5 cm and Pe < 16.

Analytical solutions to Equation 1 assume v and D to be constant with depth and time. Experimenters often carefully control the water content and thus v at the top of the column by misting or dripping the influent solution on the soil surface at some rate less than the saturated conductivity of the soil and then allow the solution to leave the column by dripping, e.g. saturated flow. The result is an uneven distribution of the water content, flow rate, solute residence time and air porosity down the length of the soil column. Under these conditions the many variables can only be separated with difficulty in order to interpret the observed results.

Induction of mass gas flow while sampling the soil atmosphere can also lead to quite spurious results. This is a particular problem in samples obtained near the soil surface, where the air above the soil may be sampled rather than the soil atmosphere at the desired depth (Starr et al., 1974).

In controlled atmosphere and water content studies, as in Starr and Parlange (1975), owing to the effect of the mass air flow on the soil atmospheric pressure, it is necessary to initiate horizontal air flow before vertical water flow. However both end plates must have a pore size of 10 - 15 µm and have been previously saturated with water to prevent air entry. Tensiometers must also be used to measure the soil water pressure. Owing to the differential air pressure on the two sides of the plates, the water pressure in the plate reservoir is not indicative of the soil water pressure. Further, the resistance to flow through the entrance plate commonly increases (K_s decreases) with time in column studies utilising N and C substrates. Hence it is necessary to have a means of maintaining a constant infiltration flux as with a positive displacement peristaltic pump.

REFERENCES

Allison, F.E. 1966. The fate of nitrogen applied to soils. Advances in Agron. 18, 219-258.

Arrhenius, E. (ed.) 1977. Nitrogen - an essential life factor and a growing environmental hazard. Nobel Symposium No. 28, Ambio. 6, 95- 182.

Biggar, J.W. and Nielsen, D.R. 1976. Spatial variability of the leaching characteristics of a field soil. Water Resour. Res. 12, 78-84.

Boast, C.W. 1973. Modelling the movement of chemicals in soils by water. Soil Sci. 115, 224-230.

Bouma, J., Zieball, W.A., Walker, W.G., Olcott., P.G., McCoy, E. and Hole, F.D. 1972. Soil adsorption of septic tank effluent. University of Wisconsin-Extension, Geological and Natural History Survey, Soil Survey Division. Info. Circ. No. 20.

Broadbent, F.E. and Clark, F. 1965. Denitrification. In: Soil Nitrogen. Agronomy, 10, 344-359. Amer. Soc. of Agron., Madison, Wis.

Corey, J.C., Nielsen D.R. and Kirkham, D. 1967. Miscible displacement of nitrate through soil columns. Soil Sci. Soc. Am. Proc. 31, 497-501.

Dancer, W.S., Peterson, L.A. and Chesters, G. 1973. Ammonification and nitrification of N as influenced by soil pH and previous N treatments. Soil Sci. Soc. Am. Proc. 37, 67-69.

Ensminger, L.E. and Pearson, R.W. 1950. Soil Nitrogen. Advances in Agronomy 2, 81-111.

Hill, D.E. and Parlange, J.-Y. 1972. Wetting front instability in layered soils. Soil Sci. Soc. Am. Proc. 36, 697-702.

Kirda, C., Nielsen, D.R. and Biggar, J.W. 1973. Simultaneous transport of chloride and water during infiltration. Soil Sci. Soc. Am. Proc. 37, 339-345.

Kirkham, D. and Powers, W.L. 1972. Advanced soil physics. p. 405. Wiley-Interscience, New York.

Macura, J. and Kunc, F. 1965. Continuous flow method in soil microbiology. V. Nitrification. Folia Microbiol. 10, 125-134.

McLaren, A.D. 1970. Temporal and vectorial reactions of nitrogen in soil: A review. Can. J. of Soil Sci. 50, 97-109.

McLaren, A.D. 1975. Comment on kinetics of nitrification and biomass of nitrifiers in a soil column. Soil Sci. Soc. Am. Proc. 39, 597-598.

Misra, C., Nielsen, D.R. and Biggar, J.W. 1974. Nitrogen transformation in soil during leaching. II. Steady state nitrification and nitrate reduction. Soil Sci. Soc. Am. Proc. 38, 294-299.

Nielsen, D.R. and Biggar, J.W. 1962. Miscible displacement. III. Theoretical considerations. Soil Sci. Soc. Am. Proc. 26, 216-221.

Nielsen, D.R., Biggar, J.W. and Erh, K.T. 1973. Spatial variability of field-measured soil-water properties. Hilgardia 42, 215-259.

Smith, M.S., Firestone, M.K. and Tiedje, J.M. 1978. The acetylene inhibition method for short-term measurement of soil denitrification and its evaluation using ^{13}N. Soil Sci. Soc. Am. Proc. (In press).

Starr, J.L. and Parlange, J.-Y. 1975. Nonlinear denitrification kinetics with continuous flow in soil columns. Soil Sci. Soc. Am. Proc. 39, 875-880.

Starr, J.L. and Parlange, J.-Y. 1976a. Relation between the kinetics of nitrogen transformation and biomass distribution in a soil column during continuous leaching. Soil Sci. Soc. Am. Proc. 40, 458-460.

Starr, J.L. and Parlange, J.-Y. 1976b. Solute dispersion in saturated soil columns. Soil Sci. 121, 364-372.

Starr, J.L. and Parlange, J.-Y. 1977. Plate induced tailing in miscible displacement experiments. Soil Sci. 124, 56-60.

Starr, J.L. and Sawhney, B.L. 1979. Movement of nitrogen and carbon from a septic system drainfield. J. Envir. Qual. (Submitted 8/78).

Starr, J.L. Broadbent, F.E. and Nielsen, D.R. 1974. Nitrogen transformatic during continuous leaching. Soil. Sci. Soc. Am. Proc. 38, 283-289.

Starr, J.L., DeRoo, H.C. Frink, C.R. and Parlange, J.-Y. 1978. Leaching characteristics of a layered field soil. Soil Sci. Soc. Am. Proc. 42, 386-391.

Takagi, S. 1960. Analysis of the vertical downward flow of water through a two-layered soil. Soil Sci. 90, 98-103.

Volz, M.G. and Starr, J.L. 1977. Nitrate dissimilation and population dynamics of denitrifying bacteria during short term continuous flow. Soil Sci. Soc. Am. Proc. 41, 891-896.

Wagenet, R.J., Biggar, J.W. and Nielsen, D.R. 1977. Tracing the transformations of urea fertiliser during leaching. Soil Sci. Soc. Am. Proc. 41, 896-902.

Warrick, A.W., Biggar, J.W. and Nielsen, D.R. 1971. Simultaneous solute and water transfer for an unsaturated soil. Water Resources Res. 7, 1216-1225.

CONVERSION AND DEGRADATION OF ORGANIC AND INORGANIC NITROGEN COMPOUNDS FROM HEAVILY LOADED WASTE WATERS IN AMPHIBIC SOILS

R. Kickuth
Gesamthochschule Kassel, 3430 Witzenhausen,
Nordbahnhofstrasse 1a, West Germany.

Under European climatic conditions most soils are able to accept by incorporation and degradation the following amounts annually of biodegradable material and nutrients:

1 025 - 2 550 kg	Organic matter/ha
60 - 140 kg	Phosphorus/ha
235 - 585 kg	Nitrogen/ha

This material may be livestock effluents, human sewage or industrial discharges. The purification capacity of soils does not exceed the equivalent of 80 - 100 people/ha and furthermore reasonable degradation rates cannot be achieved to provide a satisfactory discharge.

The two main reasons for the limited degradation capacity of terrestrial soils are the rates of microbial and chemical turnover of organic matter and nitrogen compounds in soils. The infiltration capacity restricts hydraulic loads.

Both of these processes are associated with the aerobic concept for land treatment, and this has to be maintained as a consequence of the physiological and ecological behaviour of natural or crop vegetation which depends on aerobic conditions in its root-zone. Therefore, depending on the soil type and the local rainfall, the hydraulic loads have to be restricted to 290 - 750 mm/year.

To overcome the disadvantages and limitations of aerobic degradation of carbon compounds and subsequent nitrification of the ammomium formed with its consequences for groundwater contamination, the establishment of an anaerobic limnic degradation system for purification purposes has been advantageous. The first and only condition for realising this is to abolish the agricultural uses of the irrigation plot and to grow a vegetation type, which is adapted to anaerobic conditions in the root-zone.

Swamp vegetation consisting of *Phragmites*, *Juncus*, *Typha* and *Schoenoplectus* species will fulfil these conditions. All these plants show a marked aerenchymatic tissue, by which oxygen can be transported downwards from the stem to the root. By this mechanism, not only can the oxygen demand of the root be met but also a very small aerobic zone will be formed in and adjoining the rhizosphere.

A purification plant, established in 1974 (March), has been observed over 4 years.

The irrigated area has a slight inclination of 0.4 degrees away from the inlet. The root range of *Phragmites* is restricted by the presence of a clay layer more than 1 m deep. Therefore, water does not move vertically and is transported and cleaned by horizontal movement through the root-zone of about 60 cm.

The system will deal with the following annual loads:

110 000 - 135 000 kg   Organic matter/ha
 12 000 - 14 500 kg   Nitrogen/ha
  3 900 -   4 600 kg   Phosphorus/ha

The total volume was equivalent to a hydraulic load of 13 000 mm/year.

Accumulation of mud or obstruction of flow has not so far occurred. From March 1974 to August 1978, 2 800 m^3 of suspended organic matter has been applied to the purification pond. In August 1978 the residue was estimated to be not more than 250 m^3

There are no unpleasant odours, although degradation takes place anaerobically. Apparently, either any thiols, amines and other unpleasant compounds evolved are bound instantly to the soil matrix or the system is adequately poised by the iron present. This pilot-project requires 2 m^2 for the equivalent of each head of population. This means that the plant is 50 to 60 times as effective as the land treatment following conventional aerobic treatment.

The analytical data of the wastes and purified effluents have been given elsewhere (Kickuth, 1977). In the last year the pathway and the rate of nitrogen turnover in the irrigation plot have been investigated.

All values are based on analytical results except the value for the denitrification loss. This has been obtained as the difference between the nitrogen input and the total of the amounts of nitrogen of the various compartments shown in Figure 1.

Attention is drawn to the following points:

1) A degradation rate of 98% of the effluents has been achieved.
2) Recycling of plant material formed in the course of the year returns 196 kg of nitrogen to the soil and on the basis of the cellulose content (about 20 000 kg/ha annually) leads to an additional annual incorporation of nitrogen of 1 310 kg/ha in the form of bacterial protein and humic material (13.3%).
3) Organic matter formed to a depth of 5 cm has a C : N-ratio of 9 : 1. The additional formation of N-rich organic matter on the surface of the soil raises the F-layer by 0.9 mm/year.

This shows that it is not desirable to harvest the plant material for removing nutrients.

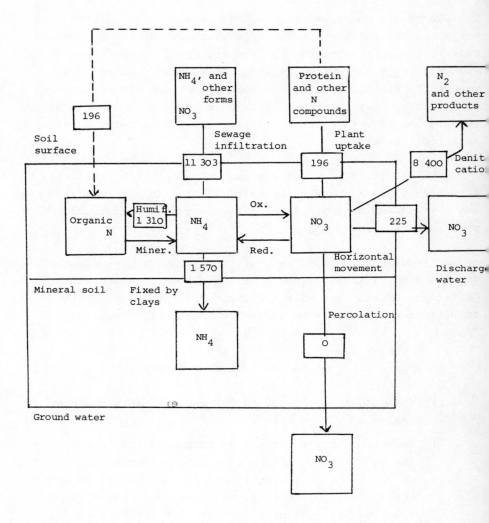

Fig. 1. Conversion and residues of the combined nitrogen of sewage by infiltration into active soil.

The values are the average for the years 1976/1977. All data are in kg related to a surface area of 1 ha.

Another 14% of the applied nitrogen fixed by the three-layer clay minerals is an important contribution to the removal of nitrogen from wastes. But this process still could be regarded as one of the limiting factors in the performance of the purification process, when expanding clay minerals are exhausted.

The outstanding process which is most effective and theoretically indefinite in operation is the loss of nitrogen by denitrification, by which more than 70% of the applied nitrogen is removed. This and the minimal concentration of NO_3^- in the discharged water are the most striking differences to the conventional aerobic processes.

REFERENCES

Kickuth, R. 1977. Degradation and incorporation of nutrients from rural waste waters by plant rhizosphere under limnic conditions. Utilisation of Manure by Land Spreading, Comm. Europ. Commun. EUR. 5672e. S. 335-342.

DISCUSSION

M.C. Cheverry *(France)*

In the experiment where fine sandy loam was underneath coarse sand, Dr. Starr observed fingers in the plot; do these correspond to a special arrangement of the soil particles, or not?

J.L. Starr *(USA)*

Where fingers occur in the field is undoubtedly influenced by old root channels and the arrangement of particles. However, for the laboratory studies we deliberately made an imperfection between the two layers and the finger would always form at that spot. Their occurrence is more related to the conductivities of the two layers and instability occurring at the interface.

A. Dam Kofoed *(Denmark)*

Dr. Starr, you spoke mostly about microbiological reduction of nitrate, what about the geochemical reduction in the deeper layers of soil? In Denmark I believe that only 35 - 40% of the rainfall passes into the drains, the rest goes into the groundwater. However, more than nine tenths of the groundwater used for drinking water is totally free from nitrogen. This suggests that the geochemical reduction of nitrate produces pure water. Have you any comments about geochemical reduction?

J.L. Starr

I am afraid that I do not have this data. However, laboratory studies show that nitrate is reduced chemically only under certain conditions, for example, very low pH. In the north eastern United States many drinking waters contain from 15 to 25 ppm nitrate nitrogen which suggests that there this kind of continued reduction of the nitrates does not occur. However, we need more measurements and a better understanding of the process.

E.A. Paul *(Canada)*
Dr. Dam Kofoed, what prevents biological reduction at depth?

A. Dam Kofoed
There is no firm evidence but soil conditions do not appear to be suitable.

T. Walsh *(Ireland)*
In practical terms, Dr. Starr, how near are you coming to a method which would enable us to determine the nitrogen status of the soil?

J.L. Starr
I think we are a long way from that yet.

H. Vetter *(West Germany)*
Dr. Kickuth, is your method for degradation of organic matter suitable only for municipal waste water or can it also be used for slurry from pigs and poultry?

R. Kickuth *(West Germany)*
We have mixed municipal waste water with about 15% of effluents from livestock, including those from a large battery installation. This increased the BOD to about 1 000 - 2 500 mg/l compared with about 500 - 600 mg/l normally found in municipal wastewater. As far as the process is concerned, the source of nitrogen or organic compounds is not important but it is advisable to separate those with different qualities.

T. Walsh
Dr. Kickuth, in your experimental work do you sow the *Phragmites* and is there any other plant, like *Juncus* for example, that is equally good?

R. Kickuth
The choice of plant depends on the problem. For example, *Juncus* is especially suitable to open the soil for infiltration,

and under favourable conditions it is green for almost the whole year whereas the *Phragmites* begins to die off in November and you have a brown field from November until late April or the beginning of May. Other plants have particular advantages.

T. Walsh

The reason I brought this up is that the current EEC project includes the identification of new work to carry on and proceed from here. It seems to me that you could combine this denitrification process with the production of cellulose.

R. Kickuth

This is right. However, at the present time the process itself works better if the material produced is left in the bed. We have considered, for example, the provisions of reeds for thatching but at present there is very little demand in Germany.

J.H. Voorburg *(The Netherlands)*

Dr. Kickuth, what was the C : N ratio of the wastewater?

R. Kickuth

The waste water had about 1 000 mg of BOD per litre, that means about 900 mg of cellulose containing 400 - 450 mg C; the total nitrogen was between 80 and 120 mg per litre. So the C : N ratio is 4 or $4\frac{1}{2}$: 1.

THE EFFECT OF NITROGEN IN PIG SLURRY, SPREAD OUT AT DIFFERENT APPLICATION TIMES

H. Vetter and G. Steffens

Landwirtschaftliche Untersuchungs- und Forschungsstelle der Landwirtschaftskammer - Weser - Ems, 2900 Oldenburg, W. Germany.

We have two series of field trials with pig slurry at our institute. In the first series, started in September 1974, 0, 30, 60 and 90 m^3/ha were spread in autumn or winter. In August 1976 we started the second series of field trials, examining the effects of different times of application of slurry on plant yield and the movement of nutrients in the soil. We want to report this second series of field trials, especially the effects of nitrogen in the pig slurry.

LOCATIONS AND WEATHER

The second series of field trials with different application times of the pig slurry has four experiments, two on sandy soils and two on clayey-silts. Table 1 gives some of the characteristics of the four soils.

The sandy soils contain about 5% clay, the clayey silts about 12%; the silt contents of the sandy soils are 1 and 5% and of the heavier soils between 36 and 56%. The sandy soils have 2 - 3% of organic C and the clay-silts between 1.5 and 2%. The nutrient status of the sites varies from medium to good.

All the sites have an impermeable soil layer at a depth of 2 to 3 m. Above this layer the shallow ground water ponds in winter.

Figure 1 shows the rainfall at the sites of the field trials on sandy soils, and Figure 2 shows the rainfall at the sites of the field trials on clayey silt. The average monthly temperatures of all the 4 locations from August 1976 to July 1978 are shown in Figure 3.

TABLE 1

SOIL CHARACTERISTICS

Treatment		Depth cm	Clay %	Silt %	% N	% C	pH CaCl$_2$	P$_2$O$_5$	K$_2$O mg/100 g	Mg	Cu ppm
1	hs	0 - 30	4.0	3.6	0.17	2.77	4.9	30	8	4.2	3.9
		30 - 50	3.0	5.5	0.09	1.62	4.6	12	5	2.5	0.7
		50 - 70	3.3	1.0	0.04	0.56	4.3	5	4	1.1	0.2
2	hs	0 - 30	4.2	6.2	0.16	2.05	4.9	28	11	4.4	5.2
		30 - 50	6.4	3.8	0.09	1.55	5.0	10	7	2.9	1.1
		50 - 70	3.0	6.9	0.04	0.62	4.9	3	6	1.9	0.3
3	tU	0 - 30	12.4	54.0	0.14	1.61	5.6	33	18	8.3	2.4
		30 - 50	12.0	55.7	0.09	1.08	5.6	15	17	6.3	2.5
		50 - 70	16.4	46.0	0.04	0.26	5.2	3	9	7.5	0.7
4	tU	0 - 30	12.0	53.7	0.16	1.80	5.0	33	22	5.0	4.2
		30 - 50	16.1	55.7	0.10	1.06	5.1	16	22	4.3	1.9
		50 - 70	16.0	36.1	0.04	0.45	4.6	4	15	4.0	0.5

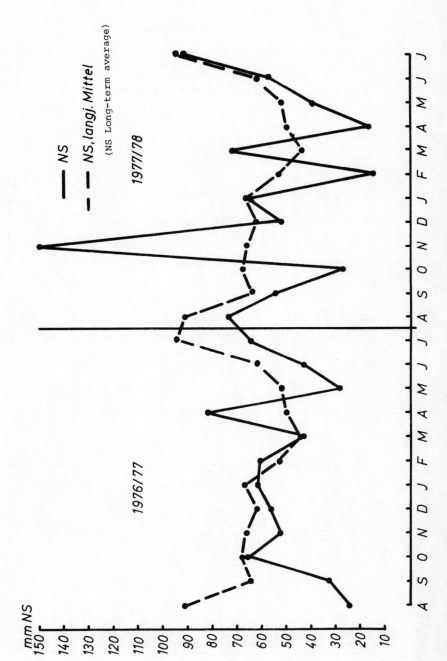

Fig. 1. Precipitation (NS) on the field-trials with humus sand, August 1976 - July 1978.

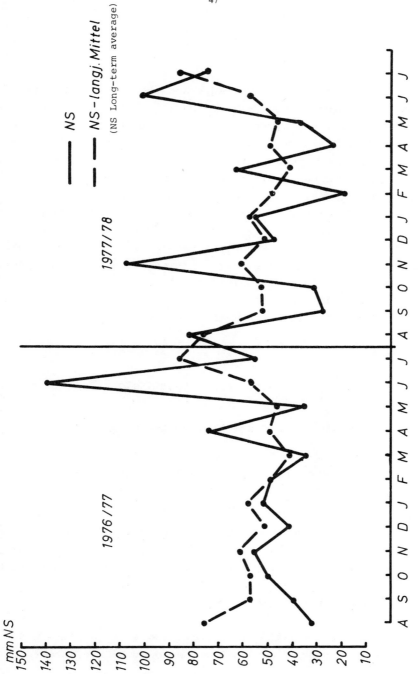

Fig. 2. Precipitation (NS) on the loamy field trials, August 1976 - July 1978.

EXPERIMENTAL LAYOUT

Table 2 gives the treatments and the average amounts of nutrient applied as slurry each year. Apart from the treatment without slurry, pig slurry was applied at the following rates and times: 30 m^3/ha in August, October, December and February/March and one treatment with 15 m^3/ha in February/March. With 30 m^3/ha, the sandy soils received 190 kg N/ha, 100 kg P_2O_5/ha and 120 kg K_2O/ha, the clayey silts received 220 kg N/ha, 180 kg P_2O_5/ha and 110 kg K_2O/ha and with 15 m^3/ha half of these amounts were applied. All the slurry treatments were replicated fourfold. The size of the plots was 150 m^2. To examine if the N-requirement of the plants were met by the slurry dressings, all the slurry plots were divided into three parts and received no addition or fertiliser -N supplying 20 and 40 kg N/ha at the beginning of the growing season.

RESULTS

We report the results of the four field trials in four parts:

1) The nitrate-and ammonium contents in the soil,
2) The total-N-contents in the soil,
3) The N-contents in the shallow ground water,
4) Yields of crops.

1) Nitrate and ammonium contents in the soil

At the beginning of crop growth, soil samples were taken from the different slurry plots. Table 3 gives the nitrate-N-contents in the soil at the end of February/beginning of March, before spreading slurry in February/March.

The upper half of Table 3 shows the average nitrate-N-contents of the two sandy soils, the lower half of the table those of the two clayey silts. In the whole soil profile from

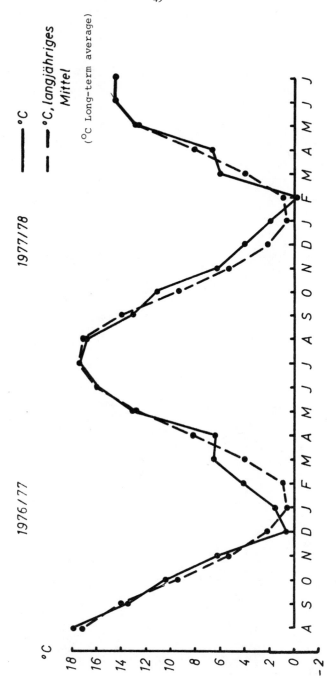

Fig. 3. Average monthly temperatures from August 1976 to July 1978.

TABLE 2

TREATMENTS AND AMOUNTS OF NUTRIENTS (kg/ha) IN THE SLURRY SPREAD IN 1977 AND 1978

Treatments		Without slurry	30 m^3 slurry August	30 m^3 slurry October	30 m^3 slurry December	30 m^3 slurry Febr./March	15 m^3 slurry Febr./March
					kg/ha		
hs	N	190	190	190	190	190	95
	P$_2$O$_5$	100	100	100	100	100	50
	K$_2$O	120	120	120	120	120	60
tU	N	220	220	220	220	220	110
	P$_2$O$_5$	180	180	180	180	180	90
	K$_2$O	110	110	110	110	110	55

0 to 90 cm deep the clayey silts contain 2 or 3 times as much
nitrate as the sandy soils. The larger amount of nitrogen in
the heavier soils corresponds to the greater water holding
capacity of these soils.

TABLE 3

NITRATE-N-CONTENTS IN THE SOIL (kg/ha) AT THE END OF FEBRUARY/BEGINNING OF
MARCH 1978, BEFORE SPREADING THE SLURRY IN FEBRUARY/MARCH

Trial			Depth			
			0 - 30 cm	30 - 60 cm	60 - 90 cm	0 - 90 cm
			kg/ha			
hs	without		12	5	5	22
∅	30 m^3	August	17	8	4	29
tr. 1	"	October	15	12	12	39
tr. 2	"	December	70	27	7	104
	"	Febr./March	14	11	4	30
	15 m^3	Febr./March	13	6	5	24
tU	without		22	27	35	84
∅	30 m^3	August	15	30	47	92
tr. 3	"	October	26	46	61	133
tr. 4	"	December	83	61	61	205
	"	Febr./March	19	33	61	113
	15 m^3	Febr./March	21	26	42	89

Also the unmanured plots of these trials had more nitrogen
on the heavier soils than on the sandy soils. The sandy soils as
well as the loamy soils were manured with slurry for some years
before starting the field trials.

The different application times of the slurry show similar
effects on both soil types. After spreading the slurry in
August, both soils have stored little nitrogen, those manured
in October contained slightly more. The plots manured in
December, contained the most nitrate-N on both the sandy and the
heavier soils. Some plots on the heavier soils, manured with
30 m^3/ha of pig slurry one year before sampling (February/March)

were found to have 30 kg N more in the profile than the unmanured plots.

A short time after spreading the slurry, the topsoil is enriched with nitrate-N. The nitrogen stored over the longer times in the heavier soils, is caused primarily by the higher storage capacity of the subsoil.

The N-contents of the soils for two of the four field trials were also determined at the end of August 1978. That was in trial 1 on the humus sand and in trial 3 on the clayey silt (Tables 4 and 6), although the slurry had been spread in August on trial 1.

TABLE 4

MINERAL-N IN THE SOIL (kg/ha) AT THE END OF AUGUST 1978 TRIAL 1. hs.

			Depth			
			0 - 30 cm	30 - 60 cm	60 - 90 cm	0 - 90 cm
			kg/ha			
NO_3^--N						
Without slurry			50	12	7	70
30 m^3 slurry		August	83*	28*	7*	118*
"	"	October	42	11	6	59
"	"	December	37	30	12	70
"	"	Febr./March	30	16	5	52
15 m^3	"	Febr./March	32	16	5	53
NH_4^+-N						
Without slurry			3	2	2	7
30 m^3 slurry		August	2*	2*	1*	5*
"	"	October	2	1	1	5
"	"	December	2	2	2	6
"	"	Febr./March	4	2	1	7
15 m^3	"	Febr./March	3	2	1	5

* The soil samples were taken after spreading the slurry in August.

TABLE 5

MINERAL-N IN THE SOIL (kg/ha) AT THE END OF AUGUST 1978, TRIAL 3, tU

			0 - 30 cm	30 - 60 cm	Depth 60 - 90 cm	0 - 90 cm
					kg/ha	
NO_3-N						
Without slurry			48	17	3	68
30 m^3 slurry	August		50	16	14	80
"	"	October	48	16	27	90
"	"	December	47	28	23	99
"	"	Febr./March	67	21	27	116
15 m^3	"	Febr./March	53	19	16	89
NH_4-N						
Without slurry			5	4	1	10
30 m^3 slurry	August		3	3	2	7
"	"	October	3	3	2	8
"	"	December	3	2	1	6
"	"	Febr./March	2	5	2	9
15 m^3	"	Febr./March	2	3	2	7

Two weeks after spreading the slurry in August, the nitrate-N in the sandy soil is increased from 0 to 30 cm and also from 30 to 60 cm deep. The ammonium-N was not. Obviously ammonium-nitrogen in the slurry was nitrified very quickly to nitrate nitrogen. The contents of nitrate-N are 5 - 40 times greater than those of the ammonium-N.

Because the slurry had not been spread in August on the heavier soil, the nitrate-N in the soil shows the nitrogen storage of the earlier slurry applications from 6 to 12 months previously. Most nitrogen was found after spreading slurry in spring. Less nitrogen was stored after spreading the slurry in December and in October than after spreading in spring, and least after spreading in August. The amount of nitrogen available for the plants will be less, the longer the time between spreading the slurry and the beginning of crop growth.

2) The total-N contents in the soil

In these field trials there were no significant differences in the total-N contents between the different treatments after two years. To examine how much of the applied nitrogen remains in the soil, the field trials must continue longer, such as 8 - 10 years.

Other investigations at our institute on some farmers' fields have shown that even large dressings of slurry continued for many years have increased the soil-N remarkably little (Table 6). We found the nitrogen content of the whole soil profile was increased by only 700 kg N, on average of 22 fields which had received slurry over the last 15 years totalling about 15 000 kg N/ha, that is about 1 000 kg N/ha annually. Calculating a removal of nitrogen in the plants of 2 000 kg N within the 15 years, a balance of 13 000 kg N remains. Of these 13 000 kg N, only 700 kg N are stored in the soil; that is only 5% of the amount applied. Ninety-five percent of the nitrogen in the slurry has been lost.

TABLE 6
N-BALANCE AFTER MANURING FOR 15 YEARS WITH LARGE AMOUNTS OF SLURRY

Additions and losses	Amount of N (kg/ha)
Applied as slurry	15 000
Removed in the plants	2 000
Remaining in the soil	700
Total accounted for	2 700
Total loss from system	12 300
Annual loss	800

3) Nitrogen-contents in the ground water

For investigating the leaching of nitrogen in the field, we put dip wells (drain filter pipes of a diameter of 5 cm) to enable samples of the ground water to be obtained. In winter we pumped the water out of these wells every 3 - 4 weeks and then sampled the newly percolated water.

Figure 4 shows the average ammonium-N + nitrate-N contents in the shallow ground water of the two field trials on humus sand. The ammonium concentration in both the shallow ground water and the soil was very low, with few values more than 5 mg N/l.

The shallow ground water of the humus sands contained between 15 and 55 mg N/l. The largest amounts of mineral-N were found after spreading the slurry in August, October and December. The concentrations were 10 - 20 mg N/l less after spreading the slurry in February/March. After loss of mineral-N in early winter, least N was found in the shallow ground water from the plots without slurry.

We found similar results on analysing the shallow ground water of the heavier soils (Figure 5). The shallow ground water from the plots without slurry also contained least mineral-N. The ground water from the plots receiving 30 and 15 m^3 pig slurry/ha in spring contained concentrations of mineral-N intermediate in value of that from plots without slurry and the plots with slurry in August, October and December. The concentrations were 20 mg/l lower when the slurry was spread in spring than when spread in autumn.

The amounts of mineral-N in the soil, and in the shallow ground water in the heavier soils were larger than in the sandy soils. This is probably an effect of larger slurry dressings before starting the field trials. In the first year of our experiments, the soils of these field-trials still contained more mineral-N than soils not having received slurry so that differences due to the different slurry treatments were not apparent. Probably there will be a further decrease of amounts of mineral-N in the next years.

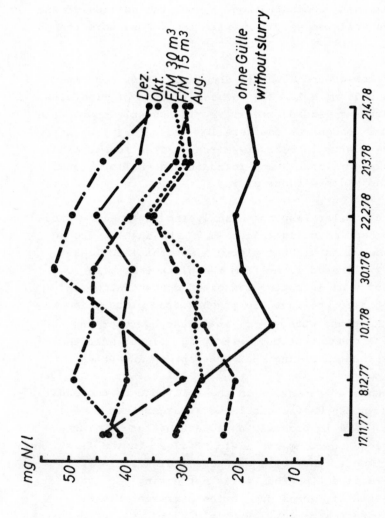

Fig. 4. N-contents in the shallow groundwater 1977/78, average of the 2 field trials on humus sand.

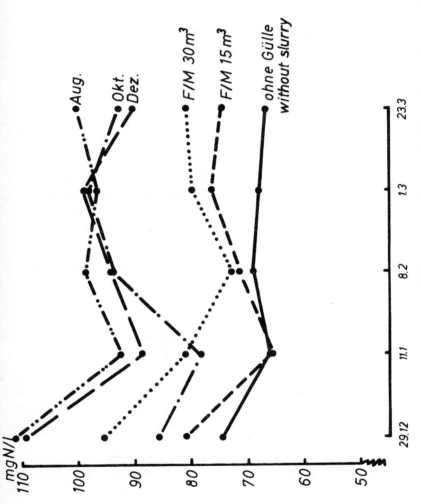

Fig. 5. N-contents in the shallow groundwater 1977/78, average of the 2 field trials on clayey silt.

In summary, on both the sandy soils and the heavier soils the shallow ground water will contain more N when slurry is spread in autumn or winter than when slurry is spread in spring

4) Yields

Figure 6 shows the effects of spreading the slurry at different times on the yield of winter cereals on the sandy soils (on average of 1977 and 1978).

a) Slurry treatments without fertiliser-N
b) Slurry treatments + 20 kg fertiliser-N
c) Slurry treatments + 40 kg fertiliser-N
   (fertiliser was applied in spring).

All three curves show that slurry-nitrogen applied in August did not increase yields of grain; there were small increases from slurry spread in October and large increases from spreading the slurry in December and in spring. By spreading 30 m^3/ha of pig slurry, containing 190 kg N/ha, in December or spring the plants had adequate N for growing in spring. Fifteen m^3 pig slurry applied in spring with 20 or 40 kg fertiliser-N/ha gave the same yield as 30 m^3 pig slurry, spread in spring. These results show that even when the slurry was spread at the best time, 95 kg slurry-N was only equivalent to 20 - 40 kg fertiliser-N.

On the heavier soils we also obtained the best yields by spreading the slurry in spring, and least yields by spreading the slurry in autumn (Figure 7). The increases in yields from slurry were smaller on the heavier soils than on the sandy soils, because unmanured cereals on the heavier soils already gave good yields of grain (5t/ha). These heavier soils had been enriched with nutrients from large dressings of slurry in the years before starting the field trials. Not only do they contain much P_2O_5 and K_2O, they also mineralise large amounts of nitrogen.

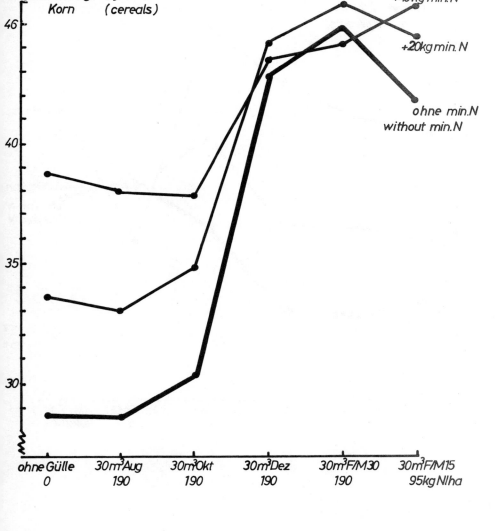

Fig. 6. Yields on humus sand (average of 1977 and 1978) after spreading the slurry at different times.

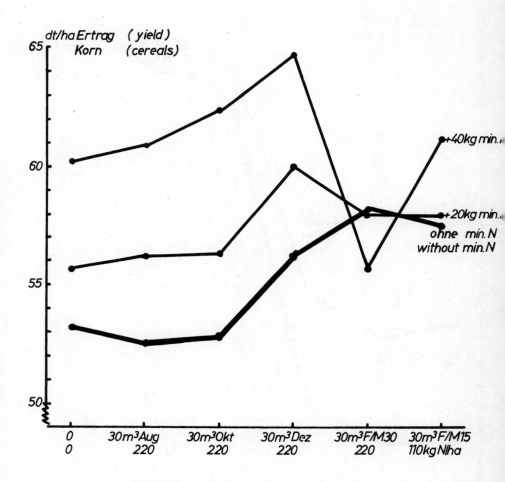

Fig. 7. Yields on clayey silt (average of 1977 and 1978), after spreading the slurry at different times.

CONCLUSIONS

When we ask what happens to the slurry-nitrogen in the soil, the following points emerge:

1. Most of the available nitrogen in the slurry will be changed in the soil very quickly into nitrate. The soil contains very little ammonium-N.
2. The heavier soils store more N than the sandy soils.
3. Very little organic nitrogen remains in the soil; also the humus content remains at the same level as in normal farming.
4. When slurry is spread in autumn, the nitrate nitrogen will leach rapidly into the deeper soil layers. Much of this nitrogen will be lost for plant growth.
5. Nitrogen losses will be greater with larger slurry dressings.
6. When we try to prepare a nitrogen balance sheet, we have the following values:

	Amounts of N kg/ha in system from slurry applied in:	
	Autumn	Spring
30 m^3 pig slurry/ha	190*	190*
Removed in the crop	0*	25*
Lost by leaching	60*	25*
Remainder (storage, volatilisation, experimental error)	130*	140*

* Nitrogen amounts deriving from the last slurry dressing.

The distribution of the remainder between its components cannot be done. Long-term experiments are needed to give valid calculations of the N-storage in soil and the N volatilisation to the air.

MATHEMATICAL MODELLING OF COMPLEX REACTIONS AND SOLUTE MOVEMENT THROUGH SOILS

J.H. Dufey

Département Science du Sol, Université Catholique de Louvain,
Place Croix du Sud 2, 1348 Louvain-la-Neuve, Belgium.

This communication shows how a simple equation of convective dispersive movement may be used to support sophisticated sub-models describing complex reactions of solutes in soils.

The vertical movement of any solute through soil profile obeys the continuity equation:

$$\frac{\partial c}{\partial t} = -\frac{1}{\phi}\frac{\partial J}{\partial x} + Q \qquad (1)$$

where c is the concentration of the solute in the soil solution ϕ is the water-filled porosity, J is the solute flux resulting from diffusion and soil solution displacement:

$$J = \phi\left(-D\frac{\partial c}{\partial x} + vc\right) \qquad (2)$$

where D is the diffusion-dispersion coefficient, and v is the average vertical rate of water movement in soil pores. At constant rate,

$$\frac{\partial c}{\partial t} = D\frac{\partial^2 c}{\partial x^2} - v\frac{\partial c}{\partial x} + Q \qquad (3)$$

The Q term is the contribution of any physical, chemical, or biological reaction to the change of solute concentration during its transfer in solution, i.e. what is produced minus what is destroyed per time unit.

The method for solving that general equation will mainly depend on the purpose of the experiment. Investigations of the dispersive aspect of solute movement have received most effort, for example the effect of substrates on dispersion coefficients In such cases, if the solute is just a means for investigating

the pore medium, a chemical species is selected for which
$Q = 0$, e.g. chloride or tritiated water. On the other hand,
experimental conditions are established so that the initial
and boundary conditions are simple enough to enable the general
equation to be solved by analytical methods. Then the work
consists of adjusting a theoretical curve to experimental points
in order to get parameters of interest of the material such as
dispersion coefficient and porosity (Laudelout and Dufey, 1977).

When the Q term has to be taken into account, analytical
solutions are possible only in a few limited and simple cases,
e.g. the movement of solutes, that are adsorbed on solid
surfaces linearly with their concentration in solution.

In natural soil conditions, the difficulty of solving the
general equation (3), by analytical methods mainly lies in
establishing the initial conditions; for example if the solute
concentration is not uniform throughout the soil profile at zero
time no analytical solution can be used. Numerical solutions
must then be used.

Many numerical methods have been developed to solve the
continuity equation with $Q = 0$. Some of these solutions agree
fairly well with the analytical solutions where these exist.
Using finite differences to give approximate values of
derivatives agreement is usually obtained at the cost of
extremely small increments of time and length. Figure 1 shows
a comparison between an implicit numerical solution and the
analytical solution (Brenner, 1962; Parlange and Starr, 1975).
For example, for P (Peclet Number) = 4, the profile length has
been divided into 50 increments and the time into 100 increments
per pore volume. This means that 100 systems of 50 equations
with 50 unknowns must be solved to simulate the percolation of
one pore volume. If the solute is not involved in any reaction,
this is of course possible; but if Q is not zero the calculations
may become prohibitive, except again in very simple and
artificial cases.

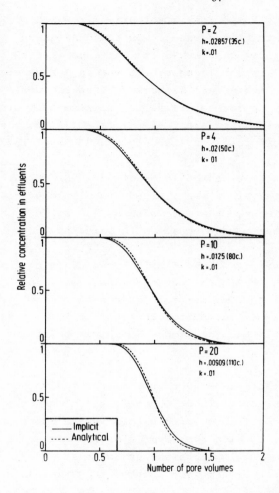

Fig. 1. A comparison between an implicit numerical solution of Equation 3 and its analytical solution, for various Peclet numbers (h: reduced increment of length; k: reduced increment of time).

A choice then has to be made. Is it worthwile to develop complex numerical solutions if one knows that they will not be able to support any realistic 'Q submodels'? Or would the use of less sophisticated, but maybe also less accurate, numerical equations of movement be better, to which very advanced 'Q submodels' could be connected? Only experience can answer that question.

A very simple solution of the continuity equation is found by using the following approximations:

$$\frac{\partial c}{\partial t} \approx \frac{1}{k} (c_{i,j} - c_{i,j-1}) \qquad (4)$$

$$\frac{\partial^2 c}{\partial x^2} \approx \frac{1}{h^2} (c_{i+1,j} - 2 c_{i,j} + c_{i-1,j}) \qquad (5)$$

$$\frac{\partial c}{\partial x} \approx \frac{1}{2h} (c_{i+1,j} - c_{i-1,j}) \qquad (6)$$

where h and k are the increments of length and time respectively; $c_{i,j}$ is the concentration at length ih and time jk. Choosing the following values for h and k

$$h = \frac{2D}{v}, \qquad k = \frac{2D}{v^2}$$

gives the solution

$$c_{i,j} = \frac{1}{2} (c_{i-1,j} + c_{i,j-1}) \qquad (7)$$

In terms of dimensionless units

$$h' = \frac{h}{L}, \qquad k' = \frac{vk}{L}$$

where L is the profile length, which means that Equation 7 is valid when

$$h' = k' = \frac{2D}{vL} = \frac{1}{2P}$$

with P (Peclet Number) = vL/4D. This requires the profile to be divided into 2P layers, as well as the time required for the percolation of one pore volume.

The same solution (7) holds true when using

$$\frac{\partial c}{\partial x} \simeq \frac{1}{h} (c_{i+1,j} - c_{i,j}) \qquad (8)$$

instead of Equation 6, provided that one chooses

$$h = \frac{D}{v}, \qquad k = \frac{D}{v^2}$$

or

$$h' = k' = \frac{1}{4P}$$

i.e. the profile length and the reduced unit time must be divided into 4P increments.

Figure 2 shows that using 4P layers results in very good agreement between the numerical and analytical solutions for P = 10, and 20. For P = 2, and 4, the coincidence is not so good (one reason being that both solutions have not been submitted to the same boundary conditions), but it is still quite satisfactory from a practical point of view.

As Equation 7 does not involve any resolution of equation systems, programming is very easy, even on desk calculators. Moreover, due to the extreme simplicity of that model of dispersive-convective movement, very elaborate submodels can be used at each step of length and time to describe the chemical, physical, or biological reactions in which the solute is involved. Up to the present we have applied that method of simulation in our studies on salt movements. The transfer of Ca, Mg, and Na salts involves complicated reactions such as ion exchange, precipitation, solubilisation, ion pair formation. We have built submodels describing those reactions (Dufey et al 1978) and they have been coupled to Equation 7 (Chattaoui et al., 1978; Belhaj et al., 1978; Goblet, 1978).

The good agreement between simulated and experimental results reported in those papers, seems to indicate that eventual imperfections of Equation 7 are masked by the experimental errors on the measurements on the one hand, and by the likely large contribution of reactions to the solute dispersion on the other hand.

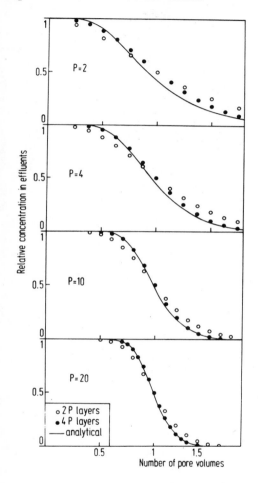

Fig. 2. A comparison between the explicit numerical solution, Equation 7, of Equation 3 and its analytical solution, for various Peclet numbers.

For nitrogen movement in soils, we have attempted to introduce into Equation 7 the submodel of ammonium oxidation presented by Laudelout et al. (1977).

$$NH_4^+ \xrightarrow[\text{Nitrosomonas}]{\frac{3}{2} O_2 \downarrow} NO_2^- \xrightarrow[\text{Nitrobacter}]{2H^+ + H_2O \uparrow} \xrightarrow{1/2 O_2 \downarrow} NO_3^-$$

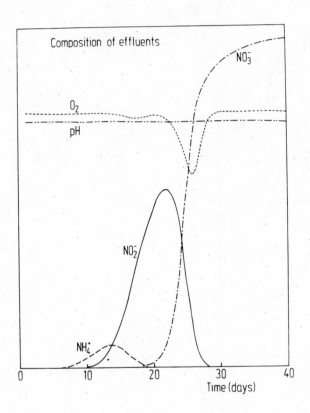

Fig. 3. Simulated nitrogen concentrations in effluents from a sand column with ammonium applied.

The main factors included in that model are the initial densities of micro-organisms, substrate concentrations, oxygen concentration, pH, and temperature. In this first attempt at simulating N movement, we have assumed that micro-organisms do not move with soil solution, and that ammonium was supplied on the surface of a sandy soil, i.e. NH_4^+ ion exchange was neglected in this first step. The simulated breakthrough curves reported in Figure 3, have shapes well in agreement with previously published results (Morrill and Dawson, 1967; Erh et al., 1967).

REFERENCES

Belhaj, Md., Gallez, A., Chiang, C. and Dufey, J.E. 1978. Simulation physique et mathématique du mouvement des sels dans les sols irrigués du Tadla. Hommes, Terre et Eaux, sous presse.

Brenner, H. 1962. The diffusion model of longitudinal mixing in beds of finite length. Numerical values. Chem. Eng. Sci. $\underline{17}$, 229-243.

Chattaoui, T., Dufey, J.E. and Laudelout, H. 1978. Simulation physique et mathématique du mouvement des sels dans les sols de la Haute-Vallée de la Medjerda de Tunisie. 'Sols de Tunisie'.

Dufey, J.E., Petit, C.M., Goblet, Y. and Laudelout, H. 1978. Modélisation des équilibres physico-chimiques d'échange et de précipitation dans les systèmes sol - eau - électrolyte. Annales Agronomiques, à paraître.

Erh, K.T., Elrick, D.E., Thomas, R.L. and Corke, C.T. 1967. Dynamics of nitrification in soils using a miscible displacement technique. Soil Sci. Soc. Am. Proc. $\underline{31}$, 585-591.

Goblet, Y. 1978. Simulation physique et mathématique du mouvement des sels dans les sols irrigués. Application aux eaux carbonatées. Mémoire Ing. Agronome, Dépt. Science du Sol, Louvain-la-Neuve.

Laudelout, H. and Dufey, J.E. 1977. Analyse numérique des expériences de lixiviation en sol homogène. Annales Agronomiques, $\underline{28}$(1), 65-73.

Laudelout, H., Lambert, R. and Pham, M.L. 1977. Mathematical modelling of ammonium oxidation in effluents. 1977. Utilisation of Manure by Landspreading, Commission of the European Communities, 447-453.

Morrill, L.G. and Dawson, J.E. 1967. Patterns observed for the oxidation of ammonium to nitrate by soil organisms. Soil Sci. Soc. Am. Proc. $\underline{31}$, 757-760.

Parlange, J.Y. and Starr, J.L. 1975. Linear dispersion in finite columns. Soil Sci. Soc. Am. Proc. $\underline{39}$, 817-820.

DISCUSSION

J.K.R. Gasser *(UK)*

Professor Vetter, from the composition of the drainage water, were you able to estimate how much of the applied dressing was in it?

H. Vetter *(West Germany)*

Nearly 30% of the nitrogen from slurry applied in the autumn was found in the drainage water. In one field experiment the slurry supplied 190 kg N/ha and in the other 220 kg N/ha.

G. Chisci *(Italy)*

At what time did you sample the soil in your experiments?

H. Vetter

In practice the most important time to determine the nitrogen level is in the spring when the vegetation is beginning to grow. We determined the nitrogen content for the first time at the end of February and beginning of March. This is the most important time because the plants are beginning to need the nutrients. The second time was the month of August when samples were taken to estimate the nitrate and ammonium content.

SECOND SESSION

Chairman: T. Walsh

THE USE OF TRACERS TO DETERMINE THE DYNAMIC NATURE OF ORGANIC MATTER

E.A. Paul and J.A. van Veen*

Department of Soil Science, University of Saskatchewan,
Saskatoon, S7N OWO, Canada.

ABSTRACT

Early experiments with ^{13}C, ^{14}C and ^{15}N established the high rate of internal cycling of soil organic matter and reintroduced the concept of an active and passive phase in soil humus turnover. Later studies confirmed non-tracer investigations indicating that the percent decomposition of added materials is relatively independent of the rates of addition but dependent on its form and composition. The initial decomposition rate, plus the stabilisation of microbial products in soil, must be taken into account when interpreting degradation of ^{14}C enriched straw, roots, microbial tissue and specific components or in carbon dating naturally occurring ^{14}C. Where initial decomposition data could be described by first order kinetics we calculated decay rate constants with and without the consideration of biosynthesis. Decay rates for laboratory systems were twice those for tropical field soils and eight times those calculated for temperate climates. The data were used in a model incorporating the concepts of microbial biosynthesis and recalcitrant and decomposable soil organic fractions which can both be physically protected. This realistically described the behaviour of soil-C in a Canadian grassland before and after cultivation.

---

* Permanent address: Association Euratom-ITAL, P.O. Box 48, Wageningen, The Netherlands.

HISTORY OF TRACER RESEARCH

The advent during the 1940's of tracer research in organic matter studies came at a time when the principles affecting plant decomposition and the effects of cultural practices on total organic matter contents had been reasonably well investigated (Greaves and Carter, 1920; Phillips et al., 1935; Guha Sircar et al., 1940). Uncertainties centred around the effects of plant decomposition (Waksman and Gerretsen, 1931; Russell, 1937) and C/N ratios (Engel, 1931; Salter, 1931; Rubins and Bear, 1942) on degradation rates and stabilisation values. The possibility of soil population turnover, with the subsequent release of N on decomposition, had been postulated (Richards and Norman, 1931). However, normal methods of residue-C and N-addition followed by mineralisation studies could not separate the various aspects of the nutrient cycles (Pinck et al., 1950). Norman (1946) in reviewing the then current status of soil microbiology said "The availability of the stable nitrogen isotope N^{15} and the carbon isotope C^{13} will make it possible to verify quantitatively the various nitrogen transformations in relation to the carbon cycle and should aid greatly in establishing the forms of nitrogen present in soil".

The Pioneers

The initial investigation utilising tracers for soil organic matter studies (Norman and Werkman, 1943) labelled soybeans by growing them on ^{15}N-nitrate. The addition of this material to the soil and the growth of a new crop showed that 26% of the organic ^{15}N applied in the tagged soybeans was recovered by the subsequent crop. Subsequent work (Broadbent and Norman, 1946; Broadbent and Bartholomew, 1948; Broadbent, 1947) established the principles for utilisation of $^{13}C - ^{15}N$ tracers. It was concluded that decomposition of an available substrate was dependent on the amount added and increased the release of native soil C and N, i.e. priming. Studies with ^{14}C also noted the increased mineralisation of soil C (Bingeman et al., 1953; Kirkham and Bartholomew, 1954). Similar increased turnover of soil organic materials were indicated with ^{15}N (Bartholomew and Hiltbold, 1952; Hiltbold et al., 1951). The reiteration of the priming process which Löhnis (1926) had postulated after experiments with green manure involved much of the earlier work on isotopic studies in soil organic matter.

Jansson (1955) carried out decomposition studies with straw incubated with ammonium nitrate in which one set of samples contained labelled ammonium and other labelled nitrate. It was then possible to trace the N from the three original sources, added ammonia, added nitrate and straw N at different stages in the decomposition process. It was concluded that equilibration was caused by a cyclic mineralisation-immobilisation turnover in which the organic straw N and ammonia N participated whereas the nitrate N was practically excluded.

Development of concepts concerning the dynamic nature of soil organic matter

Broadbent and Stojanovic (1952) and Stojanovic and Broadbent (1956) in attempting to obtain more specific information about the rate of mineralisation and immobilisation applied the equations of Kirkham and Bartholomew (1954, 1955) to their data. This study appears to be the first in which a mathematical analysis was conducted together with ^{15}N. Other studies on organic N mineralisation (Wallace and Smith, 1954) showed subsequent loss of the mineralised N by denitrification and increased utilisation of non-tagged N in the presence of ^{15}N (Walker et al., 1956): the tagged N being removed by microorganisms and replaced by non-tagged N.

Jansson (1958) defined the preferential utilisation of ammonia by micro-organisms and the great extent of mineralisation-immobilisation. He showed that the priming reaction, which caused such a controversy during the early years of tracer work, was influenced by the general cycling of the nutrients. Previous authors had suggested the possibility of different fractions of soil organic matter turning over at different rates. Jansson's cross-wise tagging and mathematical treatment utilising Kirkham and Bartholomew's 1954 and 1955 equations, established the presence of an active and passive phase. He found 10 - 15% of the N to be active. The concept of a passive and active organic phase and of ammonia being more closely related to microbial synthesis than nitrate is shown in Figure 1

Development of tracer techniques

During the 1960's experimental techniques for labelling and measurement of isotopes in soil organic matter studies were detailed (IAEA, 1966, 1968). Most labelling is conducted by $^{14}CO_2$ assimilation via the leaves in a closed system. Growth to maturity is required so that uniform labelling occurs (Chekalov and Illyuviyeva, 1962; Sauerbeck and Führ, 1966; Smith et al., 1963; Scully et al., 1956). More recently, specific components of plants have been labelled by injection of ^{14}C-precursors (Crawford et al., 1977; Martin and Haider, 1977).

Counting of enriched materials is now routinely done with scintillation equipment. The availability of techniques for conversion of large amounts of naturally occurring C into benzene for counting in a scintillation counter has made it possible for a number of laboratories to undertake carbon dating. However, gas counters for C_2H_2 and CO_2 also are still employed. Present methods for plant labelling and counting of soil organic matter as well as carbon dating have recently been described (IAEA, 1976).

INTERPRETATION OF TRACER DATA ON ORGANIC MATTER DYNAMICS

The driving force for much of the organic matter turnover in soil is the microbial search for the energy tied up in reduced C compounds. An understanding of this process requires a knowledge of:

1) The chemistry of the soil organic matter constituents,
2) Levels of input of plant and animal residues,
3) Rate of decomposition of these residues,

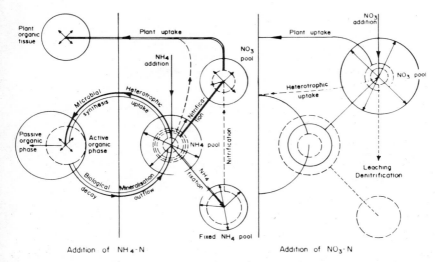

Fig. 1. Relations between the internal nitrogen cycle and addition of inorganic nitrogen to the soil. Nitrifying soil, net mineralisation conditions. (Jansson, 1958).

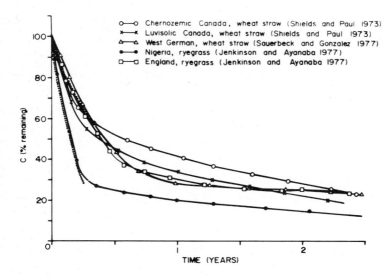

Fig. 2. Residual carbon left after field incubation of wheat straw and grass residues in different environments. The dotted lines (....) represent output of the model (Figure 9) describing decomposition of complex substrates.

4) The extent of stabilisation of microbiologically produced decomposition products, and

5) The availability of mathematical concepts and modelling techniques to help interpret the data.

The following discussion will consider each of the fractions, i.e. plant residues, root exudates, microbial biomass, specific plant and microbial components, and resistant organic components under separate headings. A complete discussion involving N and C is impossible in one review. Comprehensive reviews (Hauck and Bremner, 1976) and bibliographies (Hauck and Bystrom, 1970) have been developed for ^{15}N investigations in soils. This paper concentrates on the use of C isotopes, either ^{14}C enriched materials, ^{13}C as a natural tracer in the atmosphere or as an enrichment, and ^{14}C occurring in the atmosphere both from cosmic radiation and from the more recent bomb tests.

Plant residue decomposition

Initial work with ^{13}C (Broadbent, 1947; Broadbent and Bartholomew, 1948) concluded that small quantities of plant residues decomposed more rapidly than large quantities. Pinck and Allison (1951) summarising non-isotopic work on this subject however, concluded that the percent decomposed was nearly always independent of the quantity added if the C addition did not exceed 1.5% of the soil dry weight and if decomposition was allowed to continue for at least 3 to 6 months. The non-tracer data were later confirmed with tracers (Stotzky and Mortensen, 1958; Jenkinson, 1965, 1977; Oberländer and Roth, 1968). After extended incubation periods, the proportion of added plant material retained in various soils under different climatic conditions using different plant materials and rates of addition is very similar (Führ and Sauerbeck, 1968; Jenkinson, 1964; Sauerbeck and Gonzalez, 1977). This is often due to a similarity in production and stabilisation of soil organic matter rather than to equal rates of decomposition (Paul and McLaren, 1975). The curves in Figure 2 show a wide divergence in original decomposition rates. The slowest process is represented by the Saskatchewan grassland which is affected by drought in summer and by extended periods of frost. Sauerbeck and Gonzalez (1977) found similar rates for a number of German soils and those in Costa Rica. The data for Rothamsted are similar to those for Germany, but different for Nigeria. Jenkinson and Ayanaba (1977) found that the Rothamsted and Nigerian data could be superimposed if the Nigerian time scale was divided by a factor of four.

Attempts to relate decomposition to plant composition have led to recalculation of some of the extensive non-tracer data. The data of Waksman and Tenney (1928) have been fitted (Herman et al., 1977) to equations including carbon and nitrogen content and lignin and carbohydrate composition:

$$\frac{1}{CO_2 \text{ evolved}} = \frac{1}{CH_2O \text{ loss}} = \text{lignin-C loss} = \frac{(\frac{C}{N} \text{ straw. \% lignin})}{\sqrt{\% \text{ carbohydrates}}}$$

Hunt (1977) found that the best equation describing Pinck et al.'s (1950) data on decomposition of a wide range of plant residues to be:

$$S_O = 0.070 + 1.11 \sqrt[3]{(N/C)} ,$$

where S_O is the initial proportion of easily decomposable constituents and N/C the N to C ratio, with $r^2 = 0.98$. Both equations include the C/N ratio but the relationship of Herman et al. (1977) takes more plant residue characteristics into account.

Lespinat et al. (1976) who summarised plant decomposition studies on various crops found that different soil types with varying concentrations of clay and organic matter resulted in carbon mineralisation rates ranging from 30 to 81% for uniformly labelled plant materials added at rates ranging from 77 to 784 mg C/100 g soil. In their experiments, leaves, whole plants and water soluble extracts mixed with Oxisols, Mollisols and Endosols gave no differences in mineralised C after two weeks.

Differences, however, were significant during the first two
weeks particularly for water soluble extracts and stem and
leaves.

Pal and Broadbent (1975) found that turnover times of
uniformly labelled immature rice straw ranged between 0.8
and 3.4 years depending on loading rate and soil type.
Addition of plant material resulted in a net loss of C. Thus,
actual priming in addition to increased turnover of organic
constituents had occurred in their experiments. Immature
materials often used in laboratory studies yield different
decay values and stabilisation rates than do mature residues.

Decomposition of roots and root exudates

Roots constitute a major C reservoir in natural
ecosystems and play a major role in continuously cropped
systems where most above ground material is removed. Some
work has indicated similar rates of decomposition for above
and beneath ground materials (Jenkinson, 1971, 1977); others
have indicated slower rates for roots (Nyhan, 1975). Root
exudates are an additional source of C for microbial growth
and soil organic matter production. Martin (1977a, b) showed
that 5 to 15% of the photosynthetically fixed C was excreted to
soil by roots. In an experiment with wheat, 56% of the fixed
C remained above ground, 28% was found in the roots, 6.6% in
the soil, and rhizoshpere respiration accounted for 9.2%.

Wheat labelled at the heading stage in Saskatchewan had
52% of the labelled C in the shoots, 25% in the shoot bases
and the roots, and beneath ground respiration accounted for
23% (Warembourg and Paul, 1973). Labelling at the dough stage
resulted in 69% of the photosynthate being present above ground,
14% in the roots and shoot bases, and 17% was lost by
respiration. These data therefore are similar to that of
Martin's (1977a, b) except that in the latter case, negligible
C was found excreted to the soil matrix. Lespinat et al.
(1975a, b) found a close relationship (r = 0.94) between the
quantity of ^{14}C assimilated by the plants and that excreted by

the roots with old maize plants excreting more C than young plants. They found an average of 0.7% of the total photosynthate excreted.

Sauerbeck and Johnen (1977) found that ^{14}C measurements resulted in 20% higher recovery of root material than mechanical determinations. During their experiment, 20% of the total soil respiration was attributed to root respiration and 80% came from decomposition. The total respiration was three times as great as the root ^{14}C remaining at the end of the experiment.

Natural grassland labelled under field conditions retained 52% of the assimilated C above ground and 36% in the shoot bases and roots, with 12% being respired during and immediately subsequent to the labelling period (Warembourg and Paul, 1977). A half-life of 107 days for the C in the root-soil system indicated that more than half the residual root C was decomposed in one growing period. The work of Dahlman (1968) indicated a turnover time of 3 to 4 years for roots to grow, die and decompose. The turnover of living roots in a grassland system, therefore, varies with the percentage of fine root hairs and structural roots on the perennial plants.

Degradation of specific components
Aromatic compounds such as complexed phenolic and carboxylic acids comprise 50% of the total soil C. Complexes of amino acids account for 20%, whereas carbohydrates comprise 10 to 20%. The remaining 10 to 20% is found as a mixture including long chain fatty acids, alkanes, lower fatty acids, cell wall components such as techoic acids, nucleic acids, etc. Organic matter composition is remarkably similar from soil to soil over a broad range of soils (Schnitzer, 1977). There have been a large number of studies which determine the labelling and turnover of classical soil fractions such as fulvic acids, humic acids and humin (Jenkinson, 1971; Swift and Posner, 1977; Bailly et al., 1977). Physical separation of labelled materials associated with soil minerals also has been found useful in

organic matter dynamics studies (Paul and McGill, 1977). However, most meaningful turnover data probably come from technique based on measurement of specific chemical substances. Generally, fulvic acids contain a greater proportion of phenolic aromatics whereas humic acids contain more carboxylic acids (Flaig et al., 1975). Fulvic acids also tend to be lower in molecular weight and contain the soil carbohydrates soluble in sodium hydroxide. The humin contains the carbohydrates insoluble in sodium hydroxide (Glickert, 1972; Salfeld and Söchtig, 1977).

Simonart and Mayaudon (1961, 1962) fractionated the chemical components of plant and microbial materials and studied the degradation of these materials. They showed that the ^{14}C of most components rapidly entered the fulvic, humic and humin fractions and that proteins could be stabilised by adsorption to soil humic acids (Mayaudon and Simonart, 1958, 1959, 1963; Mayaudon, 1968). Plant pigments such as chlorophyll and B-carotene were converted primarily to non-hydrolysable humins and hydrolysable humic acids, whereas labelled cellulose, lignin, vanillin, syringaldahyde and para-hydroxybenzaldahyde were converted into hydrolysable and non-hydrolysable fractions of humic acids.

Carbohydrates

Carbohydrates are particularly useful for following the turnover of individual plant or microbial constituents and for measuring the stabilisation of newly produced products (Cortez, 1977). Individual sugars are rapidly transformed. Mayaudon (1971) using radiorespirometry over short time intervals found that the decomposition of glucose followed first order kinetics. The initial mineralisation rate was temperature dependent rising from 1.6 µg C/min/100 g soil at $4^{\circ}C$ to 10.8 µg C/min/100 g soil at $37^{\circ}C$. Also, the mineralisation rate varied from 3.1 at pH 5 to 6.7 µg C/min/100 g soil at pH 7. Micro-organisms contain different constituent monosaccharides than do plant residues. Mannose is widespread in fungi, fucose and rhamnose are frequent observed (Webley and Jones, 1971; Wagner and Tang, 1976).

In bacteria, the primary sugars are glucose, fructose, mannose, galactose, rhamnose and fucose. Xylose and arabinose are not major components. When hemicellulose containing residues are degraded, the decrease in xylose and arabinose and the production of mannose is a measure of microbial polysaccharide production and stabilisation (Table 1). Glucose was a constituent of the hemicellulose, but was produced during microbial growth. Its initial degradation and stabilisation, therefore, involved a great deal of internal cycling.

TABLE 1

ALTERATION IN SUGAR COMPOSITION DURING INCUBATION OF ^{14}C HEMICELLULOSE (Cheshire et al., 1974).

Sugar	Original		% of original			
	Amount (mg/g)	Specific activity ($\mu Ci/g$)	Time (days)			
			14	56	112	443
Rhamnose	0.8	3.1	–	44	68	60
Fucose	0.3	23.8	–	6	–	–
Arabinose	1.3	24.4	22	5.7	6.8	4.9
Xylose	2.6	128.6	19	5.1	4.4	2.3
Mannose	1.5	0.6	710	400	680	450
Galactose	1.6	8.6	54	30	11	17
Glucose	4.8	6.5	176	145	109	94
Reducing sugar as glucose	17.5					

Determination of ^{14}C activity of humic acid fractions brought about by synthesis of aromatic structures from added ^{14}C carbohydrates indicated that only 8 – 16% of the ^{14}C was found in aromatic constituents. This indicated that 92 – 84% of the residual ^{14}C from carbohydrates incorporated into humic acids was of a non-aromatic nature (Martin et al., 1974).

Aromatics

The turnover rate of aromatics is a major factor in determining soil organic matter dynamics. To study the degradation of aromatics natural ligno-celluloses containing ^{14}C in either lignin or cellulose components were prepared by feeding plants with ^{14}C phenylalanine or ^{14}C glucose (Crawford et al., 1977). Lag periods were very pronounced for lignin oxidation. Rates of CO_2 evolution for the ^{14}C cellulose component increased rapidly after short lag periods. The degradation rate for the lignin was 25% that of the associated cellulose indicating that even closely associated materials decompose at different rates.

The degradation of coniferyl alcohols, ferulic acid and caffeic acid, which have similar structures to lignin subunits, have been studied by Martin and Haider (1977). The $^{14}COOH$ group of many aromatics is removed by decarboxylation with little or no incorporation into soil organic constituents (Figure 3). This figure also shows the relative degradation and stabilisation rates for glucose, simple amino compounds and the ring ^{14}C of aromatics. Incorporation of the caffeic acid into humic-like polymers stabilised all carbons but the $^{14}COOH$ still showed preferential decarboxylation (Haider and Martin, 1975).

Martin and Haider (1977) showed that the bulk of the partially degraded coniferyl alcohol C incorporated into lignin was recovered in the humic acid fraction. Humic acid obtained by sodium hydroxide extraction and precipitation with acid, therefore, contained both humic acid derived ^{14}C and partly decomposed lignins. Hydrolysis with 6 N HCl removed 38% of the ring ^{14}C and 45 to 47% of the side chain and OCH_3 groups of the free alcohols that had been incubated in soil. This suggests that the majority of the side chain lignin fragments were present in microbial products such as proteins and polysaccharides, while more of the ring C was present in the non-hydrolysable aromatic components. Andreux et al. (1977) also found that specific labelled catechol polymers formed structures similar to residues of humic acid after acid hydrolysis. The first five days of incubation in soil saw active decarboxylation of the ring derived OCH_3 and preferential degradation of the glycine incorporated in the less stable, polymerised fraction.

The second stage was characterised by very slow mineralisation of stable polymers resulting from physico-chemical stabilisation.

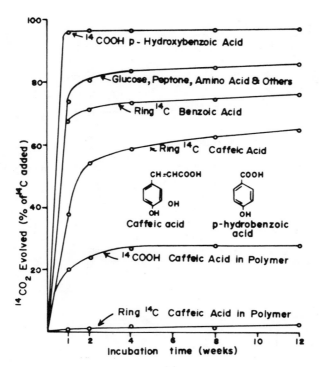

Fig. 3. Decomposition of specifically ^{14}C-labelled benzoic and caffeic acids, caffeic acid linked into phenolic polymers, and several simple organic compounds in Greenfield sandy loam. (Haider and Martin, 1975).

Amino compounds

Amino acids comprise about 20% of the soil C and 35 to 40% of the soil N but contribute a larger percentage of the potentially mineralisable N (Stewart et al., 1963a, b; Bremner, 1965). A large number of labelled amino acids are formed during microbial growth on materials such as carbohydrates (Sørensen, 1967, 1969) and acetate (Paul, 1976). An example of the production, stabilisation and subsequent degradation of amino acids and amino sugars is shown in Figure 4. Both amino acids and amino sugars were rapidly produced by day 4 at

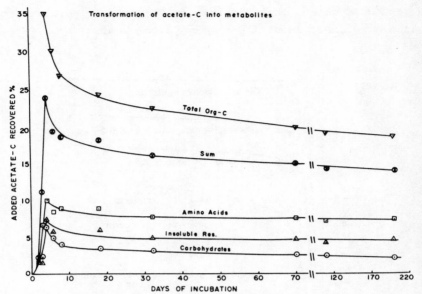

Fig. 4. Recovery of labelled carbon from soil during decomposition of ^{14}C-labelled acetate (Sørensen and Paul, 1971).

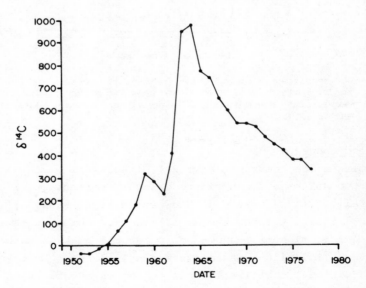

Fig. 5. Change in ^{14}C content of the wheat grown in the Northern Hemisphere (McCallum and Wittenberg, 1962, 1965; Rutherford et al., 1978).

the peak of microbial activity after the addition of ^{14}C acetate and ^{15}N ammonium sulphate. At this time, amino acids accounted for 59 μg/g labelled N out of a total of 187 μg organic labelled N/g soil, whereas amino sugars accounted for 20 μg N/g soil. At day 71, amino acids accounted for 14 and amino sugars for 9 μg N/g soil out of a total of 155 μg N/g soil remaining. The decay of these two constituents had contributed 54% of the net nitrogen mineralised between days 4 and 71. The unidentified hydrolysable fraction had increased in percentage to 50% of the total ^{15}N. Its turnover, however, accounted for only 25% of the mineralised ^{15}N. The half-life of amino sugars ^{14}C was 224 days during the latter stages of incubation (90 to 268 days). The ammonia released on hydrolysis had a half-life of 940 days and the amino acids a half-life of 2 700 days under these laboratory conditions (McGill et al., 1974).

Data similar to the above laboratory findings have also been obtained under field conditions (Legg et al., 1971) Under these conditions it was hypothesised that half of indigenous N in soil was biologically unavailable because of inaccessibility to micro-organisms of organic N compounds or complexes incorporated within aggregates. Acid hydrolysis released this N. Thus, the use of N released on acid hydrolysis results in an over-estimation of turnover rates. Nitrogen turnover rates after acid hydrolysis are best calculated on the basis of the turnover of C and calculations based on C/N ratios of the fractions before acid hydrolysis.

It is generally agreed that peptides are less sensitive to decomposition than amino acids (Haider and Martin, 1975). Although linkage of the amino acids into the synthetic polymers greatly reduces the availability to soil organisms, the amino portions of the soil organic matter molecules are more susceptible to decomposition than the ^{14}C labelled phenolic portions. In addition, data on the decomposition of ^{14}C labelled algal protein indicate that mixing with the freeze dried humic acids reduced decomposition by over 50% (Verma et al., 1975). These data substantiated Simonart and Mayaudon's (1961) conclusions that proteins were stabilised by adsorption to humic substance.

USE OF ATMOSPHERIC CARBON ISOTOPES

The two naturally occurring minor isotopes of carbon, ^{13}C which comprises 1.1%, and ^{14}C which account for 1 molecule in 10^{12} of the atmospheric-C offer a number of opportunities for measuring C dynamics when space or time limitations restrict the use of enriched materials. Humus in soils developed from plant residues which utilise the Calvin photosynthetic system has shown consistent $^{13}C/^{12}C$ ratios, with a slightly higher ^{13}C content in the humic acids than in the vegetation from which they were produced (Campbell et al., 1967) and with the humin showing the greatest discrimination ($\delta = -3.9$‰) relative to wheat (Martel and Paul, 1974). Similar uniformity of various humic fractions with regard to the ^{13}C content indicates that no gradual transformation of organic compounds from one class to another occurs, i.e. fulvic acids are not converted to humic acids (Nissenbaum and Schallinger, 1974). Scharpenseel (1977) utilised the relative ^{13}C content of soil organic matter to discount the participation of lithogenic inorganic carbonates in soil organic matter formation and to argue against the possibility of biologically inert C being present within the soil as suggested by Gerasimov (1974). The ^{13}C content of soil can be used to identify the sources of organic C in soils or sediments. Vascular plants separate into two groups relative to their ^{13}C isotope composition. Those with the C_3 (Calvin) pathway of photosynthesis have ^{13}C contents δ = -22 to -33 relative to reference 'Belemnite carbonate'. Plants which follow the C_4 (Hatch-Slack) mechanism contain carbon with δ = -10 to -20 (Bender, 1971). Haines (1976a, b), therefore, could differentiate between terrestrial C which came primarily from C_3 plants and decomposition products of Spartina which is a C_4 plant growing in estuaries. Similar comparisons on specific sites where an intermixing of known sources of C_3 and C_4 plant vegetation occurs may lead to other identifications of the specific C source in soil organic matter.

Carbon dating

Since the development of carbon dating in 1949 it has been used to investigate buried profiles in soil pedology and organic matter dynamics. Scharpenseel and Schiffmann (1977) reviewed soil carbon dating on the basis of 2 000 to 3 000 dates available for terrestrial and hydromorphic soils. They attribute the first soil carbon dates to Broecker et al. (1956) and De Vries (1958) who investigated the contaminating effects of soil constituents on fossil soils used for archeological investigations. Soil pedological investigations have included C movement and residence in podzols (Guillet, 1972; Tamm and Holmen, 1967; Gerasimov and Chichagova, 1971; Rapaire and Turenne, 1977).

The use of flotation to remove plant constituents combined with acid hydrolysis and peptisation in NaOH results in meaningful separations for carbon dating. Table 2 shows that the mean residence time (MRT) of a brown Chernozemic soil was 350 years before treatment (Martel and Paul, 1974). The fraction containing the greatest amount of ^{14}C (δ = +243‰ was the material floated on $ZnBr_2$ with a density of 2.0 g/cm^3. The C/N ratio of 17 : 1 and microscopic observation indicated that this consisted of partially decomposed plant materials. The $\delta\ ^{14}C$ for plant vegetation at this time was δ = +700 ‰ (Figure 5) indicating that the material floated by $ZnBr_2$ contained a fair amount of C originating before nuclear bomb testing.

The $\delta\ ^{14}C$ of the 0.5 N and 6 N HCl hydrolysates were +7 and +61‰, respectively. They accounted for 57% of the total C and left a residue with a $\delta\ ^{14}C$ = -197 ‰, equivalent to 1 765 mean residence time (MRT). When the soil was amended with ^{14}C acetate and incubated for 30 days, 70% of the ^{14}C added was hydrolysable compared to 61% of the ^{14}C in the radiocarbon dated field sample.

TABLE 2

MEAN RESIDENCE TIMES AND $\delta^{14}C$ OF THE FRACTIONS OF BROWN CHERNOZEMIC SOIL AND DISTRIBUTION OF ORGANIC C AND ^{14}C AS IT OCCURS IN NATURE AND 350 DAYS AFTER THE ADDITION OF LABELLED ^{14}C-ACETATE (Martel and Paul, 1974).

Fraction	Radiocarbon dated soil Ap 0 - 10 cm*					Incubated soil Ah 0 - 15 cm[+]		
	MRT	$\delta^{14}C$	^{14}C	Soil-C	^{14}C/soil-C	^{14}C	Soil-C	^{14}C/soil-C
	yr B.P. ±1σ	‰ ±σ	% of total			% of total		
Total soil	350 ± 65	-43 ± 9	100	100	1.0	100	100	1.0
Light material	Modern	+243 ± 18	8.0	6.0	0.30	5.0	5.0	1.0
ZnBr$_2$ residue	505 ± 105	-61 ± 16	97	94	0.98	95	95	1.0
0.5N hydrolysate	Modern	+7 ± 15	37	35	1.1	49	35	1.4
0.5N residue	855 ± 70	-101 ± 15	55	59	0.93	46	60	0.78
6N hydrolysate	Modern	+61 ± 22	24	22	1.2	21	20	1.1
6N residue	1 765 ± 65	-197 ± 16	31	37	0.84	25	40	0.63
NaOH extract	1 910 ± 105	-212 ± 16	22	27	0.82	16	23	0.70
Water extract	1 790 ± 120	-200 ± 15	2.6	3.1	0.84	3.4	7.0	0.48
Humin	1 330 ± 100	-153 ± 17	6.4	6.9	0.93	5.6	10.0	0.56

* Carbon content 1.9%

[+] Carbon content 2.1%

The humic acids of Chernozems tend to show the greatest MRT (Gerasimov and Chichagova, 1971; Scharpens el, 1977; Paul et al., 1964). Fulvic acids contain polysaccharides and a variety of low molecular weight microbial products with a fast turnover time. They are therefore often the youngest fraction. Table 3 shows that the fulvic acids of Eastern Canadian Gleysolic soils with a $\delta\ ^{14}C$ = +874‰ were nearly equal to new plant material ($\delta\ ^{14}C$ = +900 ‰, Figure 5). The humic acids had a δ of -140‰ equivalent to 1 220 years MRT accounting for 17% of the C. The unhydrolysed C accounting for 59% of the total C had a MRT equal to 1 530 years (-173‰). The hydrolysed C, representing 41% of the total soil C, was composed primarily of new material but had less ^{14}C than the fulvic acids.

TABLE 3

MEAN RESIDENCE TIME (MRT) OF SOIL ORGANIC MATTER FROM THE AP HORIZON OF A GLEYSOLIC SOIL (Martel and La Salle, 1977).

Soil and fractions	% of total soil C	MRT (yr B.P.±1σ)	$\sigma^{14}C$ (‰ ±1σ)
Total soil	100	Modern	+173 ± 14
Fulvic acids	24	Modern	+874 ± 64
Humic acids	17	1 220 ± 150	-140 ± 19
Humin	59	180 ± 100	-23 ± 12
Hydrolysed carbon	41	Modern	+670 ± 38
Unhydrolysed carbon	59	1 530 ± 110	-173 ± 14

Deeper soils (Scharpenseel, 1977), some podzols (Scharpenseel, 1972a, b) as well as new soils dated on a chronological sequence (Goh et al., 1976) show lower ^{14}C in the humin. The higher MRT of humin, therefore, shows less mixing with recent materials. Although fulvic acids are usually the youngest fraction, Goh et al. (1976) found that they were the oldest fractions in a soil where the movement of fulvic acid occurred. Thus, the relative MRT of the various fractions is dependent on soil pedological properties (Goh et al., 1977).

The general increase in mean residence time with depth for a large number of soils was summarised by Scharpenseel and Schiffmann (1977). The example for Chernozemic soils is shown in Figure 6. The regression equations between depth and MRT (Table 4) shows that surface soils before the influence of bomb carbon had MRT's varying from modern to approximately 1 000 years. Samples 1 to 2 meters deep increased in MRT from 5 000 to 10 000 years B.P. Contamination with bomb produced ^{14}C is not the only factor influencing the increase of MRT with depth (Herrera and Tamers, 1973). Samples obtained in 1881 from Broadbalk show a similar trend. The increase with depth must be due to differences in C input and decomposition and shows much slower turnover rates for the organic matter. Most of the soils had a slope approximating 46 yr MRT per cm depth. Soils with significantly different slopes included the Plaggen and Podzol soil which were formed under quite specific climatic conditions. The cultivated Gleysol of Canada had a buried Ah horizon at depth.

Thermonuclear bombs have created high levels of ^{14}C, nearly doubling its atmosphere content. Figure 5 shows that the lowering of ^{14}C because of industrial pollution with fossil fuels up to 1952 was rapidly overcome and reached a peak in 1964. This flush of ^{14}C is now dropping rapidly. It presents both a challenge and opportunity to the soil scientist as this ^{14}C has now entered the active phase of organic matter where it provides a sensitive universally available tracer. Goh et al. (1976) used atmospheric bomb ^{14}C to investigate the turnover of fractions in newly developing soils. Jenkinson and Rayner (1977) had data for ^{14}C plant residue decomposition over a 14-year period at Rothamsted. They also had available carbondates of the soil in 1881 and at present. The flush of bomb ^{14}C in the atmosphere was therefore used to calculate incorporation of residue C into the soil.

A large number of other laboratories have information on plant growth and C input but do not have the detailed C tracer data available at Rothamsted. These laboratories also should

TABLE 4

REGRESSION VALUES SHOWING MRT VERSUS DEPTH OF A RANGE OF SOILS WITH Y = MRT (YEARS) AND X = DEPTH

Soils	General description	Regression	(r)*	No. of samples	References
Podzols	Germany	Y = 7.47X + 1265	(.33)	32	Scharpenseel, 1972a
Udalfs	Parabrown cinnamon soils	Y = 46.51X - 4.87	(.74)	86	Scharpenseel, 1972a
Udolls	Chernozems	Y = 46.95X + 351	(.89)	122	Scharpenseel, 1972a
Plaggen	Sod soils Germany and Ireland	Y = 2.25X + 837	(.21)	34	Scharpenseel, 1972a
Vertisols	Western Europe Australia	Y = 40.14X - 421	(.77)	271	Scharpenseel, 1972a
Broadbalk 1881	England	Y = 48.9X + 695.8	(.92)	3	Jenkinson, 1969
Deciduous forest soils	Venezuela	Y = 54X - 18.21	(.96)	-	Herrera and Tamers, 1973
Eluviated Gleysol	Western Canada	Y = 67.5X + 177.4	(.94)	3	Martel and Paul, 1974

* r = correlation coefficient.

have soil samples in storage that represent well characterised sites, sampled before the advent of the hydrogen bomb in 1952. By resampling the same soils and carbon dating both the pre-bomb and the modern soils in conjunction with acid hydrolysis, it should be possible to calculate the turnover rate of the active fraction of these soils. Such data would be of great use in estimating the contribution of organic matter to soil fertility. It also is essential in understanding the effect that man is having on the global C cycle (Bolin, 1977).

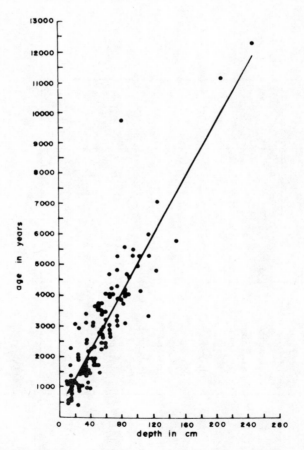

Fig. 6. Correlation between age and depth of organic matter in 122 samples from Udolls (chernozems) from the FRG, USSR, CSSR, Hungary, Bulgaria. Regression: y = 46.95x + 351.74. Correlations = 0.888 (Scharpenseel, 1972).

SOIL BIOMASS

The soil biomass has a unique position among the soil organic C fractions because it is both a sink and a transformation station for C. Biomass C does not represent a major portion of the total soil organic C, however, it is a significant part of the active phase. Jenkinson (1966) found that the biomass C was only 2.5% of the total C, but up to 12% of the ^{14}C from labelled ryegrass roots was found in the biomass after one year. In the past, the role of biomass as a sink was not readily recognised because biomass was determined primarily by plate count techniques. Plate counts account for only 1 - 15% of the bacteria identified by direct microscopy (Jensen, 1968; Babiuk and Paul, 1970). At one time it was thought the direct count over-estimated the biomass, as this technique cannot differentiate between dead and living cells (Parkinson et al., 1971). However, chemical techniques such as the measurement of CO_2 evolution after killing the biomass with chloroform followed by reinoculation and incubation (Jenkinson, 1976; Jenkinson and Powlson, 1976; Powlson and Jenkinson, 1976; Jenkinson et al., 1976) or the conversion of ATP values to biomass by the factor of 250 to 1 (Holm-Hansen, 1973; Ausmus, 1973) give still higher values.

Jenkinson (1966) found that after $CHCl_3$ sterilisation of a soil previously incubated with ^{14}C materials, the CO_2 evolved during 10 days incubation was heavily labelled. Results with experiments in which the soil was amended with ^{14}C labelled micro-organisms gave strong evidence that this heavily labelled $^{14}CO_2$ came from the biomass.

The larger estimation of biomass by measuring CO_2 evolution after fumigation with $CHCl_3$ than by the direct count of bacteria and fungi was attributed to some solubilisation of extracellular material as well as killing of organisms during treatment (Shields et al., 1974). Anderson and Domsch (1977), however, could not find any significant increase in mineralisation of prekilled fungal materials after further $CHCl_3$ treatment. The

data of Shields et al. (1974) included counts for only bacteria
and fungi. Organisms such as those with diameter larger than
3 μm make up a significant portion of the total biomass
(Jenkinson et al. 1976) although their true nature in the soil
is not known. As shown in Table 5, correcting the direct
count for these organisms significantly diminishes the difference
in biomass as measured by the two techniques. The chloroform
treatment involves a correction factor for percent of biomass
C evolved as CO_2 after chloroform treatment. The k value of
0.5 represents the data of Jenkinson and Powlson (1976), whereas
Anderson and Domsch (1977) suggest that an average of only
41% of the C is evolved as CO_2 after chloroform treatment with
the consequent k of 0.41. ATP measurements are highly affected
by the presence of clay which is known to decrease the available
P concentration. In addition to the high sensitivity of ATP
content to P availability, the inconsistency of the data of
Table 5 and the known variation of ATP in a microbial cell
during different growth stages makes this technique of biomass
measurement difficult to interpret.

  The transforming action of the biomass can be considered to
consist of three interrelated processes: uptake into the cell,
intracellular transformation, and excretion including decay of
microbial cells. When using C as a substrate, microbes take
up a fraction for biosynthesis and a fraction for energy supply.
Carbon used for energy under aerobic conditions is converted
to CO_2 and under anaerobiosis to CO_2 plus low molecular weight
organic molecules. The percentage of the total C used for
biosynthesis is found to be as low as 2.5% under anaerobic
conditions (Tusneem and Patrick, 1971) and as high as 65%
under aerobic conditions (Payne, 1970; Ladd and Paul, 1973;
Van Veen, 1977). This means that studies on the rate of CO_2
evolution in which uptake of C for biosynthesis is not considered
cannot be used for assessment of the actual decomposition rates,
but are restricted to the determination of the mineralisation
of C or the activity of microbes.

TABLE 5

COMPARISON BETWEEN BIOMASSES CALCULATED BY DIRECT COUNT, CHCl$_3$ FLUSHES AND ATP MEASUREMENT AFTER 122 HRS OF INCUBATION (Nannipieri, personal communication)

μg/g soil

	Bacterial-C from direct count after 122 h	Fungal biomass-C from direct count after 122 h	Total biomass-C by direct count after 122 h	Corrected to include large spheres $\frac{100}{64}$	Biomass calculated from CHCl$_3$ flush		Biomass from ATP after 122 h
					k = 0.5	k = 0.41	
Parent material							
Low P	21.7	102.0	123.7	198	297	362	400
High P	18.3	88.7	107.0	167	163	199	512
Parent material + clay							
Low P	27.2	93.2	120.4	189	225	274	175
High P	26.5	92.2	118.7	185	171	208	237

Biomass decomposition

Based on their biological function and chemicial composition, it is reasonable to suppose that cell walls are more resistant to biological decomposition than are cytoplasmati components(Wagner and Mutakar, 1968). After cell decay, the cell wall fraction contributes more to stable soil organic matter fraction than the cytoplasm. Hurst and Wagner (1969) showed that this is not always true since the cell wall of the fungus *Aspergillus niger* degraded faster and to a greater extent during 160 days of incubation than did the cytoplasm of this fungus. The reverse held for another fungus called S/4. They concluded that the hyphal walls of dark coloured fungal species are highly resistant to soil decomposition, and are an important contributant to soil organic matter formation. Verma and Martin (1976) did not find significant differences in the mineralisation rate of whole cells, cell walls and cytoplasm of five algae. After 22 weeks, 61 to 81% of the added C was evolved as CO_2. Decomposition of cell walls was reduced by 40% and of cytoplasm 70% after complexing cell wall and cytoplasm fractions with synthetic humic acid-type polymers, indicating the strong effect of protection by adsorption on the organic matter turnover.

Wagner and Mutaker (1968) indicate that fungi contribute more to soil organic matter formation than do bacteria. This agrees with data of Mayaudon and Simonart (1963), who found that the mineralisation of *Aspergillus niger* was less than that of *Azotobacter vinelandii*.

Maintenance Energy of the Biomass

The rate at which C is mineralised from the biomass depends on respiration for growth and maintenance of the individual cells. In pure cultures, the C requirement for maintenance is high. Barber and Lynch (1977) adapted Pirt's (1965) value of 0.04/h in their model of microbial growth in the rhizosphere. However, if this value was used for the total soil population the turnover of the biomass would be 96% per day. This is highly improbable since the biomass C is often nearly as great as the total C input from plant residues during the growing season. Behera and Wagner (1974) published data from which it could be derived that the turnover in soil was 8.6% per day. A turnover of 4.8% per day can be calculated from the soil data of Shields et al. (1973). Verstraete (1977) found that pure culture data for maintenance cannot be applied to mixed cultures. The lack of knowledge concerning the relationship between maintenance of the individual cell and maintenance of a mixed population in a complex substrate makes it difficult to extrapolate pure culture data other than for discussion purposes. Hunt (1977) applied different rates of maintenance energy to an active and inactive portion of the population. In this way the inactive population could have very low maintenance rates, whereas the actively growing organisms might approach values found for pure cultures in the literature.

DATA ANALYSIS

Pal and Broadbent (1975) found that C loss data could not be described properly by first order rate kinetics and used the form of:

$$C = kt^m$$

where C = cumulative loss of C, k = constant, t = time, m = constant. However, it is generally accepted that decomposition of organic matter constituents can be described according to first order rate kinetics (Jenkinson and Rayner, 1977; Russell, 1964; Stanford and Smith, 1972; Hunt, 1977; Gilmour et al., 1977, Sinha et al., 1977).

According to first order rate kinetics, the decomposition rate, V_{dec}, is:

$$V_{dec} = -\frac{dA}{dt} = kA$$

where k is the decomposition rate constant (time) and A is the concentration of the added organic matter. Integration of this equation gives:

$$A = A_o \cdot e^{-kt} \quad \text{or} \quad \ln \frac{A}{A_o} = -kt$$

where A_o = the concentration of A at t = 0. Values of k can be determined by plotting $\ln(A/A_o)$ vs t. In most studies, decomposition of a certain compound is determined by using ^{14}C and measuring $^{14}CO_2$ evolution or ^{14}C remaining in the soil. Neither method gives actual decomposition rates. When microorganisms utilise C, the C-compounds enter: CO_2, biosynthesis and to some extent low molecular weight metabolites. If X is the amount of CO_2-C evolved during decomposition of a certain compound and Y is the efficiency of the use of C for biosynthesis expressed as percentage of the total C uptake under aerobic conditions and non-cell metabolite production is negligible and disregarded, the actual amount decomposed is:

$$X (1 + Y/(100 - Y)).$$

A problem arising when assessing the actual decomposition rate from experimental data is the microbial production and recycling of labelled material. It is sometimes possible to determine, by chemical procedures, the amount left in soil. However, microbial production may interfere in the determination of the real decomposition rate as in the case of sugars (Cheshire et al., 1974). So ^{14}C will still be left in soil or evolved as CO_2 when the original labelled compound is completely decomposed.

Tables 6 and 7 give a number of decomposition rate constant (k) - values calculated from literature data for two values of Y, 20 and 60%, at particular times in the incubation period. Where only tabular data over extended periods were available, it was not possible to calculate k at 60% efficiency because corrections for sequential microbial growth could not be made. The gross efficiency (efficiency assuming only growth) for soil organisms under aerobic conditions is generally considered to be between 40 and 60% (Payne, 1970; Verstraete, 1977). In addition to factors such as the type of substrate other processes such as maintenance and sequential growth will decrease apparent

TABLE 6

FIRST ORDER RATE CONSTANTS CORRECTED FOR MICROBIAL BIOSYNTHESIS DURING PLANT RESIDUE AND MICROBIAL TISSUE DECOMPOSITION IN SOIL

Product	Time (day)	k/day Uncorrected	Corrected efficiency (%) 20	Corrected efficiency (%) 60	References
Plant residues:					
Straw - rye (<53 μm)	14	0.02	0.03	0.11	Cheshire et al. (1974)
Straw - rye (<1 000 μm)	14	0.02	0.02	0.04	Cheshire et al. (1974)
Straw - wheat (average for 3 soils)	28	0.02	0.03	–	Sauerbeck and Führ (1968)
Straw - wheat (average for 12 soils)	45	0.006	0.008	–	Sauerbeck and Führ (1977)
Straw - wheat	180	0.005	0.008	–	Shields and Paul (1973)
Straw - wheat	365	0.002	0.003	–	Sauerbeck and Gonzalez (1977); Oberländer and Roth (1974)
Grass - rye	5	0.07	0.09	0.24	Jenkinson (1977)
Grass - rye	10	0.04	0.06	0.23	Simonart and Mayaudon (1958)
Blue gramma herbage	30	0.006	0.008	0.02	Nyhan (1975)
Grass stipa	180	0.004	0.005	–	Shield and Paul (1973)
Roots - maize (average of 3 soils)	5	0.05	0.06	0.15	Sauerbeck et al. (1972)
Roots - maize (average of 8 values)	24	0.02	0.03	–	Cited by Lespinat et al. (1976)

(Table 6 continued)

TABLE 6 (Cont)

Product	Time (day)	k/day			References
		Uncorrected	Corrected efficiency (%)		
			20	60	
Rye tissue (54 day tissue)	10	0.02	0.02	0.06	Stotzky and Mortensen (1958)
Rye tissue (82 day tissue)	10	0.02	0.03	0.06	Stotzky and Mortensen (1958)
Microbial Tissue:					
Microcoleus sp. proteins	7	0.09	0.13	—	Verma et al. (1975)
Chlorella pyrenoidosa proteins	7	0.07	0.10	—	Verma et al. (1975)
Fungal (S/4) cell wall	10	0.02	0.03	0.07	Hurst and Wagner (1969)
Fungal (S/4) cytoplasm	10	0.04	0.05	0.17	Hurst and Wagner (1969)
A. niger cytoplasm	10	0.05	0.07	—	Hurst and Wagner (1969)
A. niger cell wall	10	0.06	0.09	—	Hurst and Wagner (1969)

efficiency. Two sequential populations of organisms each growing at 60% efficiency would have an apparent efficiency of 36%. Therefore, when using graphical data the incubation period over which the k-values in Tables 6 and 7 were calculated, was taken as short as possible.

The data in Tables 6 and 7 show that there is a wide variety in k-values for different compounds. The differences between the uncorrected k-values and the corrected ones is a factor between 0 and 2 when assuming 20% efficiency but up to a factor of 5 - 7.5 for an efficiency of 60%. This indicates that one may make a serious mistake when assessing decomposition rates of fresh amendments without accounting for the microbial biosynthesis. Another striking point is the decrease of k-values as in the case of wheat straw with time from 0.03/day at 14 days of incubation to 0.003/day at 365 days. Jenkinson and Rayner (1977) and Gilmour et al. (1977) suggested that the decomposition rate of organic matter such as straw decreased with time, because this material consists of several products such as decomposable soluble compounds, lipids, cellulose, lignin, etc., each having their own decomposition rate. K-values of single compounds, such as hemicellulose (Table 7), also decrease in time due to the earlier mentioned recycling processes.

A simple computer simulation model (Figure 7) describing the decomposition of a single compound by the amount of CO_2 evolved shows how recycling affects the apparent decomposition rate as measured. Rates of C and biomass decay follow first order rate kinetics. The rate constants of the decomposition were set for those of cellulose under laboratory conditions at 0.08/day. Estimated values for the rate constants of the decay of biomass and decomposition of microbial products were 0.05 and 0.04/day, respectively. Carbon used for maintenance energy was assumed to be 40% of the amount of C used by biomass for maintaining the population, the rest was released as microbial products. Efficiency of the use of C was assumed to be 60%.

TABLE 7

FIRST ORDER RATE CONSTANTS CORRECTED FOR MICROBIAL BIOSYNTHESIS DURING DECOMPOSITION OF SPECIFIC COMPOUNDS IN SOIL

Product	Time (day)	k/day Uncorrected	Corrected efficiency (%) 20	Corrected efficiency (%) 60	References
Glucose	1.5			2.22	Ladd and Paul (1973)
Glucose	8	0.11	0.16	-	Wagner (1968)
Glucose	10	0.11	0.19	-	Simonart and Mayaudon (1958)
Hemicellulose	10	0.08	0.11	-	Simonart and Mayaudon (1958)
Hemicellulose	14	0.03	0.04	0.11	Cheshire et al. (1974)
Hemicellulose	365	0.003	0.006	-	Minderman (1968)
Cellulose	10	0.02	0.03	0.08	Simonart and Mayaudon (1958)
Cellulose	365	0.003	0.004	-	Minderman (1968)
Lignin	365	0.001	0.002	-	Minderman (1968)
Waxes	365	0.0006	0.0008	-	Minderman (1968)
Phenols	365	0.0002	0.0003	-	Minderman (1968)
Acetate	5	0.05	0.06	0.14	Sørensen and Paul (1971)
Plant solubles	10	0.06	0.09	-	Simonart and Mayaudon (1958)
Glycine	7	0.2	0.4	-	Verma et al. (1975)
Alanine	7	0.2	0.4	-	Verma et al. (1975)
Lysine	7	0.13	0.2	-	Verma et al. (1975)
Leucine	7	0.13	0.2	-	Verma et al. (1975)
Arginine	7	0.2	0.5	-	Verma et al. (1975)
Aspartic acid	7	0.2	0.4	-	Verma et al. (1975)

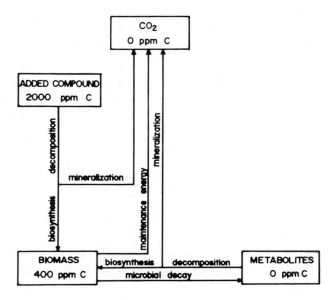

Fig. 7. Scheme of model for decomposition of a single compound (figures in the boxes refer to the initial concentrations).

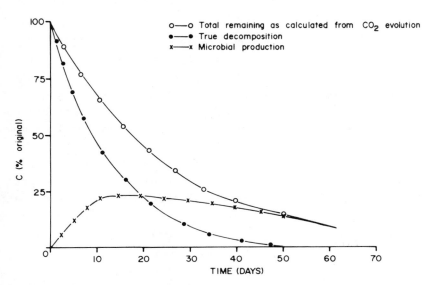

Fig. 8. Calculated decomposition of carbon added to the soil as a single compound, i.e. cellulose.

The percentage of the added C left in soil, the microbial production and the decomposition calculated from the CO_2 output are presented in Figure 8. Although the decomposition rate of biomass and the microbial products were assumed to be rather fast, accumulation of microbial products occurred. This results in a significant discrepancy between the actual decomposition and the decomposition as derived from the computed CO_2 evolution.

A second model (Figure 9) describes the decay of a natural product such as straw which consists of four components each having their own decomposition rate: soluble compounds, and proteins both with a k = 0.2/day hemicellulose k = 0.08/day and lignin k = 0.02/day (Table 7). Decay constants for biomass and microbial products as well as the data on efficiency and maintenance were equal to those of the previously mentioned model. The output (Figure 10) shows the same pattern as for a single compound except that in the latter stage the real decomposition curve levels out when only the recalcitrant fraction of lignin is left. The actual decomposition rate will differ significantly from the measured one, using CO_2 evolution data, even when differences in decomposition rate constants are taken into account.

To estimate the real decomposition rate constant from field data, we fitted our calculations to data in Figure 2. The fit between the model and the curve for decomposition of ryegrass under tropical conditions was optimal when decreasing the laboratory values of k for the four compounds considered by a factor of about 2. An eightfold decrease resulted in a good fit with the curves which represent decomposition for temperate climates. This means that the model shows a difference of a factor of four between the decomposition rate in the tropics and in temperate climates. Jenkinson and Ayanaba (1977) found that decomposition of ^{14}C residue proceeded four times as rapidly in Nigeria as in temperate England.

The half-life for the observed decomposition rate in Saskatchewan as determined from CO_2 output is 125 days. This agrees with data of Shields and Paul (1973) whose data also are

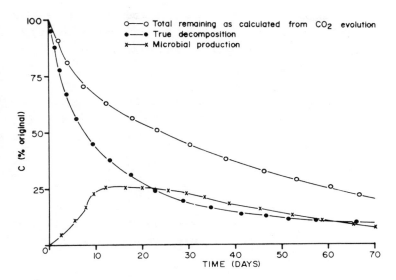

Fig. 9. Scheme of model describing decomposition of a complex substrate. (Figures in the boxes refer to initial concentrations and the figures on lines to decomposition rate constants (day).

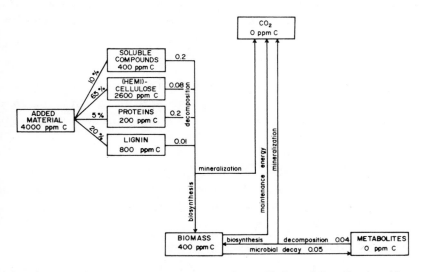

Fig. 10. Calculated decomposition of carbon added to laboratory soil as a complex substrate.

presented in Figure 2. The calculated real half-life time is 75 days. The k-values as derived from the curve in Figure 11 over the first 15 - 30 days is 0.01 per day.

Mathematical description of C dynamics in grassland

Among the many mathematical descriptions presented during the last fifteen years very few deal with soil organic C dynamics. Russell (1964) used the following expression of the rate of change of organic matter:

$$\frac{dN}{dt} = - k_1(t) N + k_2(t)$$

where $k_1(t)$ and $k_2(t)$ are the time-dependent decomposition and addition coefficients, respectively. The changes in these coefficients are represented by Fourier series which results in:

$$\frac{dN}{dt} = - (\tfrac{1}{2}\alpha_0 + \alpha_1 \cos t + \beta_1 \sin t + \alpha_2 \cos t + ...) \cdot N + (\tfrac{1}{2}\gamma_0 + \gamma_1 \cos t + \delta_1 \sin t + \gamma_s \cos t + ...)$$

Although this description results in difficult numerical solutions, it is said to be a more comprehensive representation of soil organic matter changes than the generally used equation:

$$\frac{dN}{dt} = - k_1 N + k_2$$

where k_1 and k_2 are constant in time. This equation used by Jenny (1941) to describe changes in organic N is still in common use for describing input and decomposition processes of soil organic matter (Campbell, 1978).

Jenkinson and Rayner (1977) were probably the first to publish a simulation model describing soil organic matter turn-over for a long period of time (up to 10 000 years). The data they used came from:

a) The long term Rothamsted plots of 10 - 100 years;

b) Incubation experiments during 1 - 10 years using ^{14}C-labelled plant material;

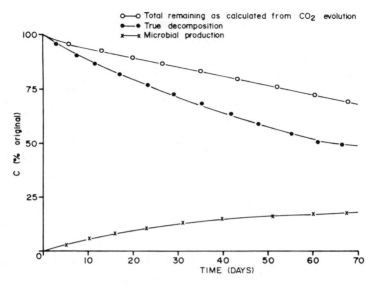

Fig. 11. Calculated decomposition of a complex substrate added to field soils representative of temperate climates.

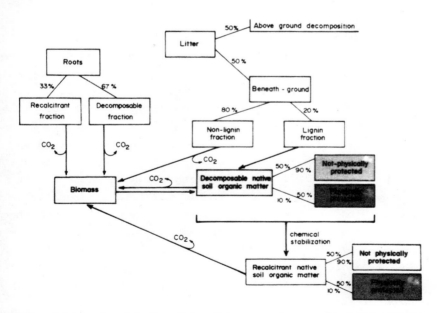

Fig. 12. Scheme of model describing C turnover in grassland and effects of cultivation.

c) Radiocarbon dating;

d) The effect of thermonuclear radiocarbon on radiocarbon age (bomb effect);

e) $CHCl_3$ data on biomass.

Five soil fractions were considered: decomposable plant material (DPM) which was calculated to have a half-life of 0.165 years; resistant plant material (RPM) with a half-life of 2.31 years; soil biomass (BIO), half-life of 1.69 years; physically stabilised organic matter (POM), half-life of 49.5 years; and chemically stabilised organic matter (COM), half-life of 1 980 years. The decomposition rate of each of these components was assumed to be proportional to its actual content. It was assumed that after decomposition any of the five components decay to the same products, CO_2, BIO, POM and COM in the same proportions. The fit between the model and experimentally determined results suggested that the model is a useful representation of the turnover of organic matter in cropped soils.

To integrate some of our own data with the wealth of information available in the literature as reviewed earlier in this paper, and to test some of our concepts concerning organic matter dynamics, we developed a computer simulation model of soil organic matter dynamics. This model was used to describe the organic matter turnover in the top 10 cm of a native grassland, and to mimic man's impact when the virgin grassland was brought into cultivation. The model is schematically presented in Figure 12 with the input data being shown in Table 8. In contrast with Jenkinson and Rayner's model, all material but the lignin fraction is transformed by passing through the biomass. While soil fauna are present and may act in increasing the turnover of biomass in this model all C is assumed to be consumed by micro-organisms, where it is either transformed into CO_2 or subsequently released by decay of microbial cells or by exudation of metabolites.

TABLE 8

INPUT DATA SIMULATION MODEL

State variables	Initial value (ppm)	Annual input (ppm)	Decay rate* constants (per day)
Litter (in soil)	780	780	0.01^+
Roots - well decomposable	4 290	666	0.02
Roots - recalcitrant	2 210	334	0.002
Biomass	500	-	0.01
Decomposable native soil organic matter - not physically protected	0^{++}	-	0.02
Decomposable native soil organic matter - physically protected	10 000	-	2.10^{-4}
Recalcitrant native soil organic matter - not physically protected	5 500	-	2.10^{-6}
Recalcitrant native soil organic matter - physically protected	5 500	-	2.10^{-6}

* Decay constants are expressed per year in the model

+ Rate constant under virgin grassland conditions is 3.0 per year and for cultivated land 3.6 per year.

++ This fraction decays very rapidly and is considered to be decayed completely at the end of a growing season when this model is assumed to start.

The lignin fraction of the incoming litter while not recalcitrant in itself was assumed to supply the majority of aromatics for soil humic materials. It enters the non-recalcitrant organic matter fraction which is also made up of microbial materials such as the amino acids and amino sugars stabilised in soil with a half-life of 5 to 20 years. In this fraction the lignin is either decomposed by the biomass or a portion can react directly with other constituents to form the recalcitrant organic matter fraction.

Organic C enters the soil via litter or roots. Approximate 50% of the above-ground litter is decomposed before it hits the soil surface (Coupland et al., 1975). The remaining 50% enters the soil to be acted on by the biomass. The proportion of root C entering the soil was calculated as 42% of the total photosynthate (Warembourg and Paul, 1973, 1977). This value includes root exudates. Earlier discussion on roots indicated that they were composed of two fractions: a resistant fraction persisting for fairly long periods, and a fraction consisting of root hairs and possibly exudates with a fairly quick turn-over rate as shown in the description of the model.

Soil organic matter consists of a number of fractions, the biomass, non-recalcitrant metabolites of the biomass, and a recalcitrant fraction. Physical protection affects the decomposition of 50% of these materials (Legg et al., 1971). The physical protection coefficient (FOPV) therefore was set at 50% for the virgin conditions. Under cultivation, this physical protection was dropped to 10%. Physical protection results from adsorption of organic C and entrapment in aggregates where it is not susceptible to micro-organisms or extracellular enzymes. In this example, physical protection of the recalcitrant material was considered not to result in a further decrease of the already very low decomposition rate constant for the recalcitrant portion of the humic materials (Campbell et al., 1967).

Transformation rates follow first order kinetics and are independent of biomass because the large biomass results in an excess concentration of organisms present at any one time in a substrate deficient system.

Very little of the extensive literature on soil organic matter turnover is directly applicable to modelling. Data on decomposition of organic matter usually yield end results, i.e. ^{14}C remaining in the soil or CO_2 evolved from the system. Model including microbial growth and metabolite production require gro in and out process data. The ^{14}C loss data in Figure 2 show the end results of a sequence of processes while the model should use individual rate constants for the processes.

The rate constants for the decomposition of the non-recalcitrant fractions of the litter and roots were obtained from Tables 6 and 7 and from calculations with the model describing plant residue turnover (Figure 11). The rate constant for the recalcitrant organic matter was derived from data on MRT (Martel and Paul, 1974). Direct data from carbon dating could not be utilised in our model. The mean residence time (MRT) (or equivalent age utilised by some other authors) is calculated direcly from the ^{14}C content of the organic matter. This value is not the same as the turnover time used in mathematical analysis. In old, slowly decomposable materials such as the humic fraction of soil organic matter, MRT and turnover time are nearly equal under equilibrium conditions. Jenkinson and Rayner (1977) calculated an equivalent age (MRT) of 2 565 years for recalcitrant organic matter and a turnover time of 2 857 years. The reciprocal of the decomposition rate constant k is equal to turnover time in first order rate kinetics. They calculated the turnover time for organisms in their model to be 2.4 years. However, the calculated radiocarbon age was 25 years because of the utilisation by the microorganisms of some C from the recalcitrant fraction that was thousands of years old.

Environmental factors such as temperature and moisture were not taken into account separately in this model. Most of the data are derived from field results which represent average values under different temperature and moisture regimes in the Canadian prairie region during the growing season. The frozen period was taken into account by a modelling programme.

The S/360CSMP (Systems/360 Continuous System Modelling Programme) (IBP Manual GH20-0240-3 and Y20-0111-0) was used as a computer language. The rectangular method of integration used intervals of 0.1.

Total soil organic matter for the virgin grassland over a 200 year period and the effect of cultivation of the grassland is given in Figure 13. On the prairie soils, man's impact is thought to result primarily from changing the fraction of the soil organic matter which is physically protected. In this model this is caused by destruction of soil structural components. This was taken into account by changing the value of the physical protection coefficient (FOPV) from 0.5 to 0.1, resulting in an increase of the decomposition rates of soil organic matter. It has been shown that disruption of soils increased the mineralisation of both soil organic C and N (Hiura et al., 1976; Rovira and Greacen, 1975; Craswell and Waring, 1972; Waring and Bremner, 1964; Edwards and Bremner, 1967). The total C input remained the same before and after cultivation. Cultivation resulted in a single large input of C in the form of the large store of grass roots present under virgin conditions. The large decrease in organic matter at the beginning of cultivation is at least in a large part due to the degradation of the roots. Although the model levels out after approximately 50 years, it does not reach steady state conditions during the period it is operable.

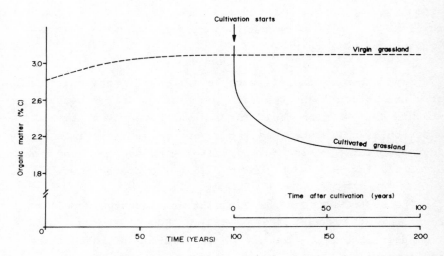

Fig. 13. Calculated organic matter vs time for virgin and cultivated grassland

The model predicted a decrease in organic C after cultivation to be 26% after 20 years cultivation. Martel and Paul (1974) found 19% loss in soil organic matter after 20 years cultivation of a Matador grassland; when the roots were taken into account the loss was 29%. The output is also very similar to other data in Canadian grasslands (Campbell et al., 1976). Soils with a higher moisture content and where cropping practices include extensive fallow periods show higher losses (Martel and Paul, 1974; Paul, 1976).

We don't consider the predictive aspect to be the most important aspect of this mathematical analysis. The analysis has allowed us to test many of the concepts derived from the review of the literature on organic C decomposition. The major points considered or arising from the analyses are:

1) Plant residues entering the soil are transformed by the biomass except for the lignin fraction.

2) The lignin fraction enters the non-recalcitrant organic C where physico-chemical reactions with microbial products can form resistant fractions. The non-recalcitrant fraction is decomposed by the biomass so that only a portion of the orignal lignin enters the recalcitrant fraction. Since microbial products also enter the recalcitrant fraction, this does not preclude microbiologically produced aromatics from forming a portion of humic materials.

3) The efficiency of the use of C by the biomass, biomass turnover, and maintenance are of major importance in calculating microbial production and organic matter dynamics. Studies utilising soil indicated that data for gross growth efficiency appears to fit the pure culture microbiological data. However, microbial turnover rates and maintenance energy concepts must be developed specifically for complex systems such as soil or sediments.

4) Physical protection and chemical recalcitrance do not result in different soil organic matter fractions as suggested by the model of Jenkinson and Rayner (1977). Our mathematical analysis assumes physical protection such as adsorption and entrapment within soil aggregates for both chemically resistant and well decomposable organic matter.

5) Growth and turnover rate data from residue decomposition or ^{14}C studies may have to be recalculated to produce values for mathematical analysis. Microbial production results in slower apparent decomposition rates, and the final amount of material remaining in soil is much more dependent on turnover of products of decomposition i.e. the recalcitrant fraction than on the decomposition rate of the plant residues added to soil.

6) The decrease in percentage of organic soil C was accurately predicted by altering the amount considered to be physically protected.

On this site there is very little difference in primary productivity between the virgin and cultivated grasslands. The major difference in plant structure is within the large reservoir of roots within the grassland. Other influences such as differences in moisture availability during non-cropped periods and differences in the time at which organic C is deposited within and on the soil surface could affect the model. These, however, will probably result in better fits to specific field data rather than in a complete change in the inputs into the mathematical system.

REFERENCES

Anderson, J.P.E. and Domsch, K.H. 1977. Mineralisation of bacteria and fungi in chloroform-fumigated soils. Soil Biol. Biochem.(in press).

Andreux, F., Golebiowska, D., Chone, T., Jacquin, F. and Metche, M. 1977. Caractérisation et transformations en milieu mull d'un modèle humique issu de l'autosydation du système catéchol-glycine et marqué sélectivement au carbone-14. In Soil Organic Matter Studies. Proc. Symp. FAO/IAEA, Sept. 1976. Braunschweig, Germany, Vol. II, pp. 43-57.

Ausmus, B.S. 1973. The use of the ATP assay in terrestrial decomposition studies. Bull. Ecol. Res. Comm. (Stockholm) 17, 223-234.

Babiuk, L.A. and Paul, E.A. 1970. The use of fluorescein isothiocyanate in the determination of the bacterial biomass of grassland soil. Can. J. Microbiol. 16, 57-62.

Bailly, J.R., Nkundikije-Desseaux, V. and Agbeko, K. 1977. Transformation d'acides phénoliques simples en substances para-humiques par des micro-organismes du sol. In Soil Organic Matter Studies. Proc. Symp. FAO/IAEA, Sept. 1976. Braunschweig, Germany, Vol. II, pp. 33-41.

Barber, D.A. and Lynch, J.M. 1977. Microbial growth in the rhizosphere. Soil Biol. Biochem. 9, 306-308.

Bartholomew, W.V. and Hiltbold, A.E. 1952. Recovery of fertiliser nitrogen by oats in the greenhouse. Soil Sci. 73, 193-202.

Behera, B. and Wagner, G.H. 1974. Microbial growth rate in glucose amended soil. Soil Sci. Soc. Amer. Proc. 38, 591-597.

Bender, M.M. 1971. Variations in the $^{13}C/^{12}C$ ratios of plants in relation to the pathway of photosynthetic carbon dioxide fixation. Phytochemistry 10, 1239-1244.

Bingeman, C.W., Varner, J.E. and Martin, W.P. 1953. The effect of the addition of organic materials on the decomposition of an organic soil. Soil Sci. Soc. Amer. Proc. 17, 34-38.

Bolin, B. 1977. Changes of land biota and their importance for the carbon cycle. Science, 196, 613-615.

Bremner, J.M. 1965. Organic nitrogen in soil. In Soil Nitrogen, Amer. Soc. Agron. Monograph 10, (Bartholomew, W.V. and Clark, C.A., eds.) pp. 93-149.

Broadbent, F.E. 1947. Nitrogen release and carbon loss from soil organic matter during decomposition of added plant materials. Soil Sci. Soc. Amer. Proc. 12, 246-249.

Broadbent, F.E. and Bartholomew, W.V. 1948. The effect of quantity of
   plant material added to soil on its rate of decomposition. Soil
   Sci. Soc. Amer. Proc. 13, 271-274.
Broadbent, F.E. and Norman, A.G. 1946. Some factors affecting the
   availability of the organic nitrogen in soil - A preliminary report.
   Soil Sci. Soc. Amer. Proc. 11, 264-267.
Broadbent, F.E. and Stojanovic, B.F. 1952. The effect of partial pressure
   of oxygen on some soil nitrogen transformations. Soil Sci. Soc.
   Amer. Proc. 16, 359-363.
Broecker, W.S., Kulp, J.L. and Tucek, C.S. 1956. Lamont natural radiocarbon
   measurements 111. Science 124, 154-165.
Campbell, C.A. 1978. Soil Organic Carbon, Nitrogen and Fertility, Marcel
   Dekker, New York. (In press).
Campbell, C.A., Paul, E.A. and McGill, W.B. 1976. Effect of cultivation
   and cropping on the amounts and forms of soil N. In Proc. Western
   Canada Nitrogen Symposium. Alberta Soil Science Workshop, Jan.
   1976, pp. 9-101.
Campbell, C.A., Paul, E.A., Rennie, D.A. and McCallum, K.J. 1967.
   Factors affecting the accuracy of the carbon-dating method in soil
   humus studies. Soil Sci. 104, 81-85.
Chekalov, K.I. and Illyuviyeva, V.P. 1962. Use of the C^{14} isotope for the
   study of the decomposition of organic matter in soil. Soviet Soil
   Sci. 1962: 482-490.
Cheshire, M.V., Mundie, C.M. and Shepherd, H. 1974. Transformation of
   sugars when rye hemicellulose labelled with ^{14}C decomposes in soil.
   J. Soil Sci. 25, 90-98.
Cortez, J. 1977. Biodégradation, in vitro, de deux lipopolysaccharides
   bactériens ^{14}C dans un sol rouge méditerranéen. Geoderma 18, 177-192.
Coupland, R.T., Willard, J.R., Ripley, E.A. and Randell, R.L. 1975.
   The Matador Project. In Energy Flow - Its Biological Dimensions. The
   IBP in Canada. (Cameron, T.W.M. and Billingsley, L.W., eds.) Publ.
   by the Canadian Committee for the IBP, Ottawa, pp. 19-50.
Craswell, E.T. and Waring, S.A. 1972. Effect of grinding on the
   decomposition of soil organic matter. II. Oxygen uptake and nitrogen
   mineralisation in virgin and cultivated cracking clay soils. Soil
   Biol. Biochem. 4, 435-442.

Crawford, D.L., Crawford, R.L. and Pometto III, A.L. 1977. Preparation of specifically labelled ^{14}C-(lignin)- and ^{14}C-(cellulose)-lignocelluloses and their decomposition by the microflora of soil. Appl. Environ. Microbiol. 33, 1247-1251.

Dahlman, R.C. 1968. Root production and turnover of carbon in the root-soil matrix of a grassland ecosystem. Int. Symp. on Methods of Productivity Studies in Root Systems and Rhizosphere, USSR, pp. 11-21.

Edwards, A.P. and Bremner, J.M. 1967. Microaggregates in soils. J. Soil Sci. 18, 64-73.

Engel, H. 1931. Über den Einfluss des C:N-Verhältnisses in verschiedenen organischen Substanzen auf die Umsetzungen des Stickstoffes im Boden. Z. Pflanzenernähr., Düng. Bodenk. 19, 314-325.

Flaig, W., Beutelspacher, H. and Rietz, E. 1975. Chemical composition and physical properties of humic substances. In Soil Components, Vol. 1, Organic Components (Gieseking, J.E. ed.) Springer-Verlag, New York, pp. 1-211.

Führ, F. and Sauerbeck, D. 1968. Decomposition of wheat straw in the field as influenced by cropping and rotation. In Isotopes and Radiation in Soil Organic Matter Studies, Technical Meeting, FAO/IAEA, Vienna, pp. 241-250.

Gerasimov, I.P. 1974. The age of recent soils. Geoderma 12, 17-25.

Gerasimov, I.P. and Chichagova, O.A. 1971. Some problems in the radiocarbon dating of soil. Soviet Soil Sci., 3-11.

Gilmour, C.M., Broadbent, F.E. and Beck, S.M. 1977. Recycling of carbon and nitrogen through land disposal of various wastes. In Soils for Management of Organic Water and Waste Waters. Proc. Symp., Publ. by ASA, CSSA, SSSA, Madison, Wis., pp. 173-194.

Goh, K.M., Rafter, T.A., Stout, J.D. and Walker, T.W. 1976. The accumulation of soil organic matter and its carbon isotope content in a chronosequence of soils developed on aeolian sand in New Zealand. J. Soil Sci. 27, 89-100.

Goh, K.M., Stout, J.D. and Rafter, T.A. 1977. Radiocarbon enrichment of soil organic matter fractions in New Zealand soils. Soil Sci. 123, 385-391.

Greaves, J.E. and Carter, E.G. 1920. Influence of moisture on the bacterial activities of the soil. Soil Sci. 10, 361-387.

Glückert, A. 1972. Note sur l'extraction et la caractérisation de polysaccharides-^{14}C formés dans le sol au cours de la biodégradation de végétaux marqués. Bulletin de l'E.N.S.A.I.A. de Nancy, pp. 69-73.

Guha Sircar, S.S., De., S.C. and Bhowmick, H.D. 1940. Microbiological decomposition of plant materials. 1. Changes in the consitituents of rice straw (Kanektara) produced by micro-organisms present in soil suspension under aerobic, anaerobic and waterlogged conditions. Ind. J. Agric. Sci. 10, 119-151.

Guillet, B. 1972. Datation des sols par le ^{14}C naturel. 11. Application à la détermination et à la signification des âges d'horizons Bh and Bs de podzols vosgiens. Bulletin de l'E.N.S.A.I.A. de Nancy, XIV/1, pp. 123-131.

Haider, K. and Martin, J.P. 1975. Decomposition of specifically carbon-^{14} labelled benzoic and cinnamic acid derivatives in soil. Soil Sci. Soc. Amer. Proc. 39, 657-662.

Haines, E.B. 1976a. Relation between the stable carbon isotope composition of fiddler crabs, plants and soils in a salt marsh. Limnol. Oceanogr. 21, 880-883.

Haines, E.B. 1976b. Stable carbon isotope ratios in the biota, soils and tidal water of a Georgia salt marsh. Est. Coastal Mar. Sci. 4, 609-616

Hauck, R.D. and Bremner, J.M. 1976. Use of tracers for soil and fertiliser nitrogen research. Adv. Agron. 28, 219-266.

Hauck, R.D. and Bystrom, M. 1970. ^{15}N. A selected bibliography for agricultural scientists. The Iowa State University Press, Iowa. 206 p.

Herman, W.A., McGill, W.B. and Dormaar, J.F. 1977. Effects of initial chemical composition on decomposition of roots of three grass species. Can. J. Soil Sci. 57, 205-215.

Herrera, R. and Tamers, M.A. 1973. Effect of vertical contamination on radio-carbon dating of soils. Acta Cient. Venezolana 24, 156-160.

Hiltbold, A.E., Bartholomew, W.V. and Werkman, C.H. 1951. The use of tracer techniques in the simultaneous measurement of mineralisation and immobilisation of nitrogen in soil. Soil Sci. Soc. Amer. Proc. 15, 166-172.

Hiura, K., Hattori, T. and Furusaka, C. 1976. Bacteriological studies on the mineralisation of organic nitrogen in paddy soils. 1. Effect of mechanical disruption of soils on ammonification and bacterial number. Soil Sci. Pl. Nutr. 22, 459-465.

Holm-Hensen, O. 1973. The use of ATP determinations in ecological studies. Bull. Ecol. Res. Comm. (Stockholm) 17, 215-222.

Hunt, H.W. 1977. A simulation model for decomposition in grasslands. Ecology 58, 469-484.

Hurst, H.M. and Wagner, G.H. 1969. Decomposition of ^{14}C-labelled cell wall and cytoplasmic fractions from hyaline and melanic fungi. Soil Sci. Soc. Amer. Proc. 33, 707-711.

International Atomic Energy Agency, 1966. FAO/IAEA Technical meeting on the use of isotopes in soil organic matter studies, Sept. 1963. Braunschweig-Völkenrode, Germany. Pergamon Press, Oxford.

International Atomic Energy Agency. 1968. FAO/IAEA Technical meeting on isotopes and radiation in soil organic matter studies. Pergamon Press, Oxford.

International Atomic Energy Agency. 1976. Tracer manual on crops and soils. FAO/IAEA Technical Reports Series No. 171. 278 p.

Jansson, S.L. 1955. Orientierende Studien über den Stickstoffkreislauf im Boden mit Hilfe von ^{15}N als Leitisotop. Z. Pflanzenernähr., Düng. Bodenk. 69, 190-198.

Jansson, S.L. 1958. Tracer studies on nitrogen transformations in soil with special attention to mineralisation-immobilisation relationships. Kungl. Lantbrukshögkolans Annaler 24, 101-361.

Jenkinson, D.S. 1964. Decomposition of labelled plant material in soil. In Experimental Pedology, Butterworths, London, pp. 199-207.

Jenkinson, D.S. 1965. Studies on the decomposition of plant material in soil. 1. Losses of carbon from ^{14}C-labelled ryegrass incubated with soil in the field. J. Soil Sci. 16, 104-115.

Jenkinson, D.S. 1966. Studies on the decomposition of plant material in soil. 11. Partial sterilisation of soil and the soil biomass. J. Soil Sci. 17, 280-302.

Jenkinson, D.S. 1969. Radiocarbon dating of soil organic matter. Rep. Rothamsted Exp. Stn for 1968, Pt 1, 73.

Jenkinson, D.S. 1971. Studies on the decompositon of C^{14} labelled organic matter in soil. Soil Sci. III, 64-70.

Jenkinson, D.S. 1976. The effects of biocidal treatments on metabolism in soil. IV. The decomposition of fumigated organisms in soil. Soil Biol. Biochem. 8, 203-208.

Jenkinson, D.S. 1977. Studies on the decomposition of plant material in soil IV. The effect of rate of addition. J. Soil Sci. 28, 417-423.

Jenkinson, D.S. and Ayanaba, A. 1977. Decomposition of carbon-^{14} labelled plant material under tropical conditions. Soil Sci. Soc. Amer. J. 41, 912-915.

Jenkinson, D.S. and Powlson, D.S. 1976. The effects of biocidal treatments on metabolism in soil. 1. Fumigation with chloroform. Soil. Biol. Biochem. 8, 167-177.

Jenkinson, D.S., Powlson, D.S. and Wedderburn, R.W.M. 1976. The effects of biocidal treatments on metabolism in soil. III. The relationship between soil biovolume measured by optical microscopy and the flush of decomposition caused by fumigation. Soil Biol. Biochem. 8, 189-202.

Jenkinson, D.S. and Rayner, J.H. 1977. The turnover of soil organic matter in some of the Rothamsted classical experiments. Soil Sci. 123, 298-305.

Jenny, H. 1941. Factors of soil Formation. McGraw Hill Book Co., New York.

Jensen, V. 1968. The plate count technique. In The Ecology of soil Bacteria, (Gray, T.R.G. and Parkinson, D. eds.) Int. Symp. University of Toronto Press, pp. 159-170.

Kirkham, D. and Bartholomew, W.V. 1954. Equations following nutrient transformations in soil, utilising tracer data. Soil Sci. Soc. Amer. Proc. 18, 33-34.

Kirkham, D. and Bartholomew, W.V. 1955. Equations for following nutrient transformation in soil, utilising tracer data II. Soil Sci. Soc. Amer. Proc. 19, 189-192.

Ladd, J.N. and Paul, E.A. 1973. Changes in enzymic activity and distribution of acid-soluble amino acid nitrogen in soil during nitrogen immobilisation and mineralisation. Soil Biol. Biochem. 5, 825-840.

Legg, J.O., Chichester, F.W., Stanford, G. and DeMar, W.H. 1971. Incorporation of ^{15}N-tagged mineral nitrogen into stable forms of soil organic nitrogen. Soil Sci. Soc. Amer. Proc. 35, 273-276.

Lespinat, P.A., André, M. and Boureau, M. 1975a. Étude quantitative du carbon libéré par les racines de mais cultivé en conditions contrôlées sous $^{14}CO_2$. Physiol. Veg. 13, 137-151.

Lespinat, P.A., Glückert, A. and Breisch, H. 1975b. The incorporation of plant root exudates in soil organic matter. In Studies about humus. Trans. Int. Symp. Humus et Planta VI, Aug. 1975. Prague, pp. 169-174.

Lespinat, P.A., Hetier, J.-M., Thomann, C., Chone, T. and Dimon. A. 1976. Utilisation de mais mûr uniformément marqué au ^{14}C pour l'étude de la matière organique de trois sols (Andosols, sol brun, sol ferrallitique) Science du Sol, Bulletin de l'A.F.E.S. 1, 53-66.

Löhnis, F. 1926. Nitrogen availability of green manure. Soil. Sci. 22, 253-290.

Martel, Y.A. and Lasalle, P. 1977. Radiocarbon dating of organic matter from a cultivated topsoil in eastern Canada. Can. J. Soil Sci. 57, 375-377.

Martel, Y.A. and Paul, E.A. 1974. The use of radiocarbon dating of organic matter in the study of soil genesis. Soil Sci. Soc. Amer. Proc. 38, 501-506.

Martin, J.K. 1977a. The chemical nature of the carbon-14 labelled organic matter released into soil from growing wheat roots. Effects of soil micro-organisms. In Soil Organic Matter Studies, Proc. Symp. FAO/IAEA, Sept. 1976. Braunschweig, Germany, Vol. 1, pp. 197-203.

Martin, J.K. 1977b. Factors influencing the loss of organic carbon from wheat roots. Soil Biol. Biochem. 9, 1-7.

Martin, J.P. and Haider, K. 1977. Decomposition in soil of specifically ^{14}C-labelled DHP and corn stalk lignins, model humic acid-type polymers and coniferyl alcohols. In Soil Organic Matter Studies, Proc. Symp. FAO/IAEA, Sept, 1976. Braunschweig, Germany, Vol. II. pp. 23-32.

Martin, J.P., Haider, K., Farmer, W.J. and Fustec-Mathon, E. 1974. Decompositon and distribution of residual activity of some ^{14}C-microbial polysaccharides and cells, glucose, cellulose and wheat straw in soil. Soil Biol. Biochem. 6, 221-230.

Mayaudon, J. 1968. Stabilisation biologique des protéines-^{14}C dans le sol. In Isotopes and Radiation in Soil Organic Matter Studies, Technical Meeting, FAO/IAEA, Vienna, pp. 177-188.

Mayaudon, J. 1971. Use of radiorespirometry in soil microbiology and biochemistry. In Soil Biochemistry, Vol. 2. (McLaren, A.D. and Skujins, J.J. eds.) Marcel Dekker, Inc., New York. pp. 202-256.

Mayaudon, J. and Simonart, P. 1958. Étude de la décomposition de la matière organique dans le sol au moyen de carbon radioactif. II. Plant and Soil 9, 381-384.

Mayaudon, J. and Simonart, P. 1959. Étude de la décomposition de la matière organique dans le sol au moyen de carbon radioactif. III. Décomposition des substrances solubles dialysables, des protéines et des hemicelluloses. Plant and Soil, II, 170-175.

Mayaudon, J. and Simonart, P. 1963. Humification des micro-organisms marqués par ^{14}C dans le sol. Ann. Inst. Pasteur, 105, 257-266.

McCallum, K.J. and Wittenberg, J. 1962. University of Saskatchewan radiocarbon dates III. Radiocarbon, 4, 71-80.

McCallum, K.J. and Wittenberg, J. 1965. University of Saskatchewan radiocarbon dating IV. Radiocarbon, 17, 229-235.

McGill, W.B., Paul, E.A. and Sørensen, L.H. 1974. The role of microbial metabolites in the dynamics of soil nitrogen. Matador Projects Tech. Rept. No. 46, April 1974.

Minderman, G. 1968. Addition, decomposition and accumulation of organic matter in forests. J. Ecol. 56, 355-362.

Nissenbaum, A. and Schallinger, K.M. 1974. The distribution of the stable carbon isotope ($^{13}C/^{12}C$) in fractions of soil organic matter. Geoderma II, 137-145.

Norman, A.G. 1946. Recent advances in soil microbiology. Soil Sci. Soc. Amer. Proc. II, 9-15.

Norman, A.G. and Werkman, C.H. 1943. The use of the nitrogen isotope N^{15} in determining nitrogen recovery from plant materials decomposing in soil. J. Amer. Soc. Agron. 35, 1023-1025.

Nyhan, J.W. 1975. Decomposition of carbon-14 labelled plant materials in a grassland soil under field conditions. Soil Sci. Soc. Amer. Proc. 39, 643-648.

Oberländer, H.E. and Roth, K. 1968. Transformation of ^{14}C-labelled plant material in soils under field conditions. In Isotopes and Radiation in Soil Organic Matter Studies. Technical Meeting, FAO/IAEA, Vienna, pp. 351-361.

Oberländer, H.E. and Roth, K. 1974. Ein Kleinfeldversuch über den Abbau und die Humifizierung von ^{14}C-markiertem Stroh and Stallmist. Die Bodenkultur 25, 111-129.

Pal, D. and Broadbent, F.E. 1975. Kinetics of rice straw decomposition in soils. J. Environ. Qual. 4, 256-260.

Parkinson, D., Gray, T.R.G. and Williams, S.T. 1971. Methods for studying the ecology of soil micro-organisms. IBP Handbook No. 19, Blackwell Scientific Publications. 116 p.

Paul, E.A. 1976. Nitrogen cycling in terrestrial ecosystems. In Environmental Biogeochemistry, Vol. 1. (Nriagu, J.O., ed.) Ann Arbor Science Publ. Inc., Ann Arbor, Mich., pp. 225-243.

Paul, E.A., Campbell, C.A., Rennie, D.A. and McCallum, K.J. 1964. Investigations of the dynamics of soil humus utilising carbon dating techniques. Trans. 8th Int. Congr. Soil Sci., Bucharest, III, 201-208.

Paul, E.A. and McGill, W.B. 1977. Turnover of microbial biomass, plant residues and soil humic constituents under field conditions. In Soil Organic Matter Studies, Proc. Symp. FAO/IAEA, Sept. 1976. Braunschweig, Germany, Vol. 1. pp. 149-157.

Paul, E.A. and McLaren, A.D. 1975. Biochemistry of the soil subsystem. In Soil Biochemistry, Vol. 3. (Paul, E.A. and McLaren, A.D., eds.), Marcel Dekker, Inc., New York, pp. 1-36.

Payne, W.J. 1970. Energy yields and growth of heterotrophs. Ann. Rev. Microbiol. 24, 17-52.

Phillips, M., Goss, M.J., Beavers, E.A. and James, L.H. 1935. The microbiological decomposition of the constituents of alfalfa hay. J. Agric. Res. 50, 761-775.

Pinck, L.A. and Allison, F.E. 1951. Maintenance of soil organic matter. III. Influence of green manures on the release of native soil carbon. Soil Sci. 71, 67-75.

Pinck, L.A., Allison, F.E. and Sherman, M.S. 1950. Maintenance of soil organic matter. II. Losses of carbon and nitrogen from young and mature plant materials during decomposition in soil. Soil Sci. 69, 391-401.

Pirt, S.J. 1965. The maintenance energy of bacteria in growing cultures. Proc. Roy. Soc. B 163, 224-231.

Powlson, D.S. and Jenkinson, D.S. 1976. The effects of biocidal treatments on metabolism in soil. II. Gamma irradiation, autoclaving, air-drying and fumigation. Soil Biol. Biochem. 8, 179-188.

Rapaire, J.L. and Turenne, J.F. 1977. Mesures d'activité specifique de fractions de matière organique appliquées a l'étude de l'évolution des sols de Guyane. In Soil Organic Matter Studies, Proc. Symp. FAO/IAEA, Sept. 1976, Braunschweig, Germany, Vol. II, pp. 179-186.

Richards, E.H. and Norman, A.G. 1931. The biological decomposition of plant materials. V. Some factors determining the quantity of nitrogen immobilised during decomposition. Biochem. J. 25, 1769-1778.

Rovira, A.D. and Greacen, E.L. 1975. The effect of aggregate disruption on the activity of micro-ogranisms in the soils. Aust. J. Agr. Res. 8, 659-673.

Rubins, E.J. and Bear, F.E. 1942. Carbon-nitrogen ratios in organic fertilisers in relation to the availability of their nitrogen. Soil Sci. 54, 411-423.

Russell, E.J. 1937. Soil Conditions and Plant Growth. 7th edition, London.

Russell, J.S. 1964. Mathematical expression of seasonal changes in soil organic matter. Nature 204, 161-162.

Rutherford, A., Wittenberg, J. and Wilmeth, R. 1978. University of Saskatchewan radiocarbon dates V. Radiocarbon (in press).

Salfeld, J.-Chr. and Söchtig, H. 1977. Composition of the soil organic matter system depending on soil type and land use. In Soil Organic Matter Studies, Proc. Symp. FAO/IAEA, Sept. 1976. Braunschweig, Germany, Vol 1, pp. 227-235.

Salter, F.J. 1931. The carbon-nitrogen ratio in relation to the accumulatio of organic matter in soils. Soil Sci. 31, 413-430.

Sauerbeck, D. and Führ, F. 1966. Experiences on labelling whole plants with C^{14}. In the use of Isotopes in Soil Organic Matter Studies, FAO/IAEA Technical Meeting, Sept. 1963. Braunschweig-Völkenrode, Germany, pp. 391-399.

Sauerbeck, D. and Führ, F. 1968. Alkali extraction and fractionation of labelled plant material before and after decomposition - a contributio to the technical problems in humification studies. In Isotopes and Radiation in Soil Organic Matter Studies, Technical Meeting, FAO/IAEA, Vienna, pp. 3-11.

Sauerbeck, D. and Gonzalez, M.A. 1977. Field decomposition of carbon-14-labelled plant residues in various soils of the Federal Republic of Germany and Costa Rica. In Soil Organic Matter Studies. Proc. Symp. FAO/IAEA, Sept. 1976, Braunschweig, Germany, Vol. 1. pp. 159-170.

Sauerbeck, D. and Johnen, B.G. 1977. Root formation and decomposition during plant growth. In Soil Organic Matter Studies, Proc. Symp. FAO/IAEA, Sept. 1976. Braunschweig, Germany, Vol. 1. pp. 141-148.

Sauerbeck, D., Johnen, B. and Massen, G.G. 1972. Der Abbau von ^{14}C-markiertem Pflanzenmaterial in verschiedenen Böden. Agrochimica 16, 62-76.

Scharpenseel, H.W. 1972a. Natural radiocarbon measurements on humic
substances in the light of carbon cycle estimates. Proc. Int. Mat.
Humic Substances. Nieuwersluis, 1972, pp. 281-292.

Scharpenseel, H.W. 1972b. Messung der natürlichen C-14 Konzentration in
der organischen Substanz von rezenten Boden. Eine Zwischenbilanz.
Z. Pflanzenernähr. Bodenk. 133, 241-262.

Scharpenseel, H.W. 1977. The search for biologically inert and lithogenic
carbon in recent soil organic matter. In Soil Organic Matter
Studies, Proc. Symp. FAO/IAEA, Sept. 1976, Braunschweig, Germany,
Vol. II, pp. 193-200.

Scharpenseel, H.W. and Schiffmann, H. 1977. Radiocarbon dating of soils,
a review. Z. Pflanzenerähr. Bodenk. 140, 159-174.

Schnitzer, M. 1977. Recent findings on the characterisation of humic
substances extracted from widely differing climatic zones. In Soil
Organic Matter Studies, Proc. Symp. FAO/IAEA, Sept. 1976. Braunschweig,
Germany, Vol. II, pp 117-132.

Scully, N.J., Chorney, W., Kostal, G., Watanabe, R., Shok, J. and
Glattfield, J.W. 1956. Biosynthesis in C^{14}-labelled plants; their
use in agricultural and biological research. Int. Conf. Peaceful
Uses of Atomic Energy, 1956, United Nations, New York, pp. 377-385.

Shields, J.A. and Paul, E.A. 1973. Decomposition of ^{14}C-labelled plant
material under field conditions. Can. J. Soil Sci. 53, 297-306.

Shields, J.A., Paul, E.A. and Lowe, W.E. 1974. Factors influencing the
stability of labelled microbial materials in soils. Soil Biol.
Biochem. 6, 31-37.

Shields, J.A., Paul, E.A., Lowe, W.E. and Parkinson, D. 1973. Turnover of
microbial tissue in soil under field conditions. Soil Biol Biochem.
5, 753-764.

Simonart, P. and Mayaudon, J. 1958. Étude de la décomposition de la
matière organique dans le sol, au moyen de carbone radioactif. 1.
Cinetique de l'oxydation en CO_2 de divers substrats radioactifs.
Plant and Soil 9, 367-375.

Simonart, P. and Mayaudon, J. 1961. Humification des protéines-C^{14} dans
le sol. 2nd Int. Symp. Pedologie, Gent, Belgium, pp. 91-103.

Simonart, P. and Mayaudon, J. 1962. Décomposition de matière organique
^{14}C dans le sol. C.R. Rech. Inst. Encour. Rech. Sci. Ind. Agric.
28, 107-112.

Simonart, P. and Mayaudon, J. 1966. Étude des transformations de la
   matière organique du sol au moyen du carbon-14. In The Use of
   Isotopes in Soil Organic Matter Studies, Technical Meeting, FAO/IAEA,
   Sept. 1963. Braunschweig-Völkenrode, Germany, pp. 245-258.

Sinha, M.K., Sinha, D.P. and Sinha, H. 1977. Organic matter transformations
   in soils. V. Kinetics of carbon and nitrogen mineralisation in soils
   amended with different organic materials. Plant and Soil, $\underline{46}$, 579-590.

Smith, J.H., Allison, F.E. and Mullins, J.F. 1963. A biosynthesis chamber
   for producing plants labelled with carbon-14. Atompraxis, $\underline{9}$, 73-75.

Sørensen, L.H. 1967. Duration of amino acid metabolites formed in soils
   during decomposition of carbohydrates. Soil Sci. $\underline{67}$, 234-241.

Sørensen, L.H. 1969. Fixation of enzyme protein in soil by the clay
   mineral montmorillonite. Experientia $\underline{25}$, 20-21.

Sørensen, L.H. and Paul, E.A. 1971. Transformation of acetate carbon into
   carbohydrate and amino acid metabolites during decomposition in soil.
   Soil Biol. Biochem. $\underline{3}$, 173-180.

Stanford, G. and Smith, S.J. 1972. Nitrogen mineralisation potentials of
   soils. Soil Sci. Soc Amer. Proc. $\underline{36}$, 465-472.

Stewart, B.A., Porter, L.K. and Johnson, D.D. 1963a. Immobilisation and
   mineralisation of nitrogen in several organic fractions of soil.
   Soil Sci. Soc. Amer. Proc. $\underline{27}$, 302-304.

Stewart, B.A., Porter, L.K. and Johnson, D.D. 1963b. The availability of
   fertiliser nitrogen immobilised during decomposition of straw. Soil
   Sci. Soc. Amer. Proc. $\underline{27}$, 656-659.

Stojanovic, B.J. and Broadbent, F.E. 1956. Immobilisation and mineralisation
   of nitrogen during decompostion of plant residues in soil. Soil Sci.
   Amer. Proc. $\underline{20}$, 213-218.

Stotzky, G. and Mortensen, J.L. 1958. Effect of addition level and
   maturity of rye tissue on the decomposition of a muck soil. Soil
   Sci. Soc. Amer. Proc. $\underline{22}$, 521-524.

Swift, R.S. and Posner, A.M. 1977. Humification of plant materials,
   properties of humic acid extracts. In Soil Organic Matter Studies.
   Symp. FAO/IAEA, Sept. 1976. Braunschweig, Germany, Vol. 1. pp.
   171-182.

Tamm, C.O. and Holmen, H. 1967. Some remarks on soil organic matter turnover
   in Swedish podzol profiles. Meddelelser fra Det Norske
   Skogforsøksvesen. Nr 85, pp. 69-88.

Tusneem, M.E. and Patrick, W.H. 1971. Nitrogen transformations in waterlogged
   soils. Bull. Louisiana Agric. Expt. Stn. No. 657, 75 p.

Veen, J.A. van. 1977. The behaviour of nitrogen in soil. A computer simulation model. Ph.D. Thesis, V.U., Amsterdam.

Verma, L. and Martin, J.P. 1976. Decomposition of algal cells and components and their stabilisation through complexing with model humic acid-type phenolic polymers. Soil Biol. Biochem. 8, 85-90.

Verma, L., Martin, J.P. and Haider, K. 1975. Decomposition of carbon-14 labelled proteins, peptides and amino acids; free and complexed with humic polymers. Soil Sci. Soc. Amer. Proc. 39, 279-284.

Verstraete, W. 1977. Fundamentele studie van de opbouw-en omzettings processen in microbiële gemeenschappen. Thesis, R.U. Gent. Belgium, 444 p.

Vries, H. de. 1958. Significance of microbial tissues to soil organic matter. In Isotopes and Radiation in Soil Organic Matter Studies, Technical Meeting, FAO/IAEA, Vienna, pp. 197-205.

Wagner, G.H. 1968. Significance of microbial tissues to soil organic matter. In Isotopes and Radiation in Soil Organic Matter Studies. Technical Meeting, FAO/IAEA, Vienna, pp. 197-205.

Wagner, G.H. and Mutakar, V.V. 1968. Amino components of soil organic matter formed during humification of ^{14}C-glucose. Soil Sci. Soc. Amer. Proc. 32, 683-686.

Wagner, G.H. and Tang, D.E. 1976. Soil polysaccharides synthesised during decomposition of glucose and dextran and determined by ^{14}C labelling. Soil Sci. 121, 222-226.

Waksman, S.A. and Gerretsen, F.C. 1931. Influence of temperature and moisture upon the nature and extent of decomposition of plant residues by micro-organisms. Ecology 12, 33-60.

Waksman, S.A. and Tenney, F.G. 1928. Composition of natural organic materials and their decomposition in the soil. III. The influence of nature of plant upon the rapidity of its decomposition. Soil Sci. 26, 155-171.

Walker, T.W., Adams, A.F.R. and Orchiston, H.D. 1956. Fate of labelled nitrate and ammonium nitrogen when applied to grass and clover grown separately and together. Soil Sci. 81, 339-351.

Wallace, A. and Smith, R.L. 1954. Nitrogen interchange during decomposition of orange and avocado tree residues in soil. Soil Sci. 78, 231-242.

Warembourg, F.R. and Paul, E.A. 1973. The use of $C^{14}O_2$ canopy techniques for measuring carbon transfer through the plant-soil system. Plant and Soil 38, 331-345.

Warembourg, F.R. and Paul, E.A. 1977. Seasonal transfers of assimilated ^{14}C in grassland: plant production and turnover, soil and plant respiration. Soil Biol. Biochem. 9, 295-301.

Waring, S.A. and Bremner, J.M. 1964. Effect of soil mesh size on the estimation of mineralisable nitrogen in soils. Nature 202, 1141.

Webley, D.M. and Jones, D. 1971. Biological transformation of microbial residues in soil. In Soil Biochemistry, Vol. 2. (McLaren, A.D. and Skujins, J.J. eds.), Marcel Dekker, Inc., New York, pp. 446-485.

MATHEMATICAL MODELLING OF NITROGEN TRANSFORMATIONS IN SOIL

J.A. van Veen and M.J. Frissel
Institute for Atomic Sciences in Agriculture, Wageningen,
The Netherlands.

ABSTRACT

To obtain a generally applicable model useful for different types of soil and cultivation practices in north western Europe, the fundamentals of several processes of the N-cycle in soil are described.

The first version of the model, published in 1973, included mineralisation, immobilisation, nitrification and transport of water, heat and nitrate.

In the second version, the description of the N-cycle in soil is extended by including other processes, with the emphasis on the effect of organic matter on the activity of the microbes involved. This version is divided into sub-models, each describing a particular process of the N-cycle. Sub-models are nitrification, mineralisation and immbolisation, volatilisation of ammonia, fixation of ammonium on clay minerals and transport of nitrate, water and heat. Nitrification is considered to be the result of the activity of two chemoautotrophic bacteria, Nitrosomonas and Nitrobacter. In the sub-model of denitrification the behaviour of oxygen in non-waterlogged and waterlogged soils is described to determine the occurrence of anaerobic sites where denitrification is supposed to occur. The rate of consumption of oxygen and the reduction of nitrate are considered to be related to the decomposition rate of organic carbon. Volatilisation of ammonia is considered to be a purely physico-chemical process, which depends on temperature, moisture, pH, cation exchange capacity and the ammonium concentration in soil. The sub-model of fixation of ammonium on clay minerals is based on a description of the fixation of potassium in soil. An equilibrium is assumed between free, i.e. exchangeabl. ammonium and ammonium in solution, and fixed ammonium with the fixation rate greatly exceeding the release rate. The main part of the model is the mineralisation and immobilisation sub-model, in which the effect of the transformation of carbon on the N-cycle is described. In the second version of the model, the effect of freshly added organic matter is particularly emphasised. This organic matter is divided into four components - proteins, carbohydrates, (hemi)-cellulose and lignin - so effecting differences in ease of decomposition and availability as substrate for the heterotrophic biomass. The effect of environmental factors such as temperature and moisture on the biological processes is expressed by multiplying the maximum rate with reduction factors, which have values between 0 and 1 (optimal conditions).

This model was compared with results from a combined greenhouse/field experiment. In this experiment, straw (1%), nitrogen (75 kg/ha) were added and nitrate and the biomass were measured during 200 days. The fit between the simulated and experimentally determined course of the nitrate concentration was satisfactory; the fit for the biomass content was not so good. The main reason for the discrepancy between simulation and experiment was thought to be due to an insufficient description of the role of native soil organic matter. Therefore, a third version of the mineralisation and immobilisation sub-model was developed in which both fresh and native soil organic matter are divided into components with differences in availability as substrate for microbes. Results of testing this model, with results from the greenhouse/field experiment and from plots in the newly reclaimed IJsselmeerpolders show this version to be an improvement compared with the second one.

During the last decade many mathematical simulation models for the behaviour of nitrogen in soil have been developed (Tanji and Gupta, 1978), varying from simple analytical descriptions of a single process to very complex computer models describing the total N-cycle in soil, each being characterised by the purpose for which it was developed. The model described here, based on earlier work (Beek and Frissel, 1973), was developed to study the quantitative aspects of cycling of N in the biosphere to allow accurate predictions on soil-N behaviour to be made for the various soils and cultural practices in north western Europe. The first version of the model described nitrification, mineralisation, immobilisation and transport of NO_3^-, water and heat through the soil, from experience with which the second version (Figure 1) was divided into sub-models, each describing a particular process.

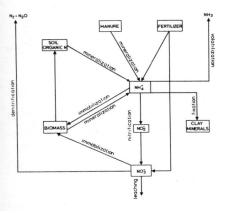

Fig. 1. Scheme of the second version of the N-model.

The use of interchangeable sub-models in the development of a complex model such as this appears very efficient provided the sub-models are compatible (Frissel et al., 1976).

This paper presents a description of several sub-models, as well as some results of testing the model with experimental data.

MODEL DESCRIPTION

The computer programme is written in CSMP, Continuous Simulation Modelling Programme. Like other specific simulation languages such as SYMPCOMP and DYNAMO, CSMP allows the correct simulation of dynamic systems such as the nitrogen cycle in soil. For simulating such systems, it is necessary to be able to calculate independently rates of change which depend on variables of state and driving variables, in any order at any instant of time. Changes are effected by integration of the rates of change over a short time interval. CSMP provides a variety of different algorithms for numerical integration. In the present model, use is made of the rectilinear method, the most simple way of integration, in which the new value of a state variable at time t + Δt equals the old value at time t plus the calculated rate of change at time t multiplied by the constant time interval Δt, which here is 0.01 day.

Environmental factors, such as temperature and moisture content of the soil, are taken into account using reduction factors. These factors, with values between 0 and 1, are calculated by interpolation from curves such as the ones shown in Figures 2a and 2b for the effect of temperature and moisture content on the soil organic matter decomposition rate.

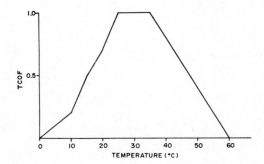

Fig. 2a. Reduction factor for the effect of temperature on the decomposition rate of soil organic matter.

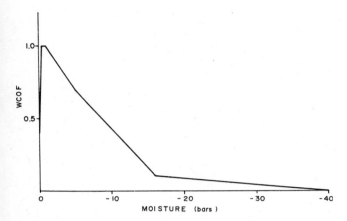

Fig. 2b. Reduction factor for the moisture effect on the decomposition rate of soil organic matter.

Data on the separate influence of environmental factors may normally be obtained with difficulty. Data on the simultaneous effect of these factors are lacking, and the programming of the mutual action is therefore more or less a guess. If the reduction factors for temperature, moisture content and oxygen pressure are represented respectively by TCOF, WCOF and O2COF, many possibilities exist for combining them. An extremely simple assumption is that all factors act independently, so that the net effect is obtained by multiplying all factors expressed in both FORTRAN and CSMP by:

$$EFFECT = TCOF * WCOF * O2COF \qquad (1)$$

If the values of the reduction factors are respectively 0.8, 0.7 and 0.4, then EFFECT = 0.224. Another simple view is that the minimum value controls the reaction:

$$EFFECT = AMIN1 \ (TCOF, \ WCOF, \ O2COF) \qquad (2)$$

and giving EFFECT = 0.4.

A more balanced view is perhaps to consider temperature separately and multiply it by the minimum value of the other two reduction factors, realised by:

$$EFFECT = TCOF * AMIN1 \ (WCOF, O2COF) \tag{3}$$

and giving EFFECT = 0.32.

We have not yet solved this problem; it will require a thorough literature research and, if that fails, extensive experiments.

The next section gives detailed descriptions of the sub-models of nitrification, volatilisation of ammonia, ammonium fixation and denitrification. The equations in the tables are written in the same form as those being used in the computer model. The values of the constants used are averages of data from the literature and are used for model calculations. More detailed information on the sub-models, as well as on the results of model calculations, were given by Van Veen (1977) and on the transport sub-model by Beek and Frissel (1973) and Frissel and Reiniger (1974).

NITRIFICATION

Nitrification is considered to be the result of the activity of two genera of bacteria *Nitrosomonas*, which converts NH_4^+ into NO_2^-, and *Nitrobacter*, which transforms NO_2^- into NO_3^-. Heterotrophic nitrification is disregarded.

The growth rates of both bacteria, GRA and GRN are described with Monod-type kinetics as shown in equations (4) and (5) (Table 1). For calculating the transformation rates of NH_4^+ and NO_2^- due to nitrification, DNH4N and DNO2, it is assumed that nitrogen only is utilised as the energy supply of the nitrifiers, described by RONH4 and RONO2 respectively.

TABLE 1

EQUATIONS OF THE SUB-MODEL OF NITRIFICATION

$GRA = REGRAM * AMOX * TCOFN * AMIN1\ (O2COFN,\ WCOFN) *$ $NH4/(KNH4 + NH4)$	(4)
$GRN = REGRNM * NO2OX * TCOFN * AMIN1\ (O2COFN,\ WCOFN) *$ $NO2/(KNO2 + NO2)$	(5)
$RONH4 = GRA/YNH4$	(6)
$RONO2 = GRN/YNO2$	(7)
$DNH4N = -RONH4$	(8)
$DNO2 = RONH4 - RONO2$	(9)
$DNO3N = RONO2$	(10)
$DAMOX = GRA - KDN * AMOX$	(11)
$DNO2OX = GRN - KDN * NO2OX$	(12)

GRA = gross growth rate of *Nitrosomonas* (cell/g soil/day)
REGRAM = specific gross growth rate of *Nitrosomonas* (1.5/day)
AMOX = *Nitrosomonas* content in soil (cell/g soil)
TCOFN = temperature reduction factor for nitrification
O2COFN = oxygen conc. reduction factor for nitrification
WCOFN = moisture content reduction factor for nitrification
NH4 = NH_4^+ conc. in soil (mg N/g soil)
KNH4 = saturation constant (0.005 mg N/g soil)
GRN = gross growth rate of *Nitrobacter* (cell/g soil/day)
REGRNM = specific gross growth rate of *Nitrobacter* (1.3/day)
NO2OX = *Nitrobacter* content in soil (cell/g soil)
NO2 = NO_2^- conc. in soil (mg N/g soil)
KNO2 = saturation constant (0.01 mg N/g soil)
RONH4 = consumption rate of NH_4^+ for energy supply (mg N/g soil/day)
YNH4 = gross growth yield of *Nitrosomonas* ($4.5.10^8$ cells/mg N)
RONO2 = consumption rate of NO_2^- for energy supply (mg N/g soil/day)
YNO2 = gross growth yield of *Nitrobacter* (1.10 cells/mg N)
DNH4N = rate of change of NH_4^+ conc. due to nitrification (mg N/g soil/day)
DNO2 = rate of change of NO_2^- conc. (mg N/g soil/day)
DNO3N = rate of change of NO_3^- conc. due to nitrification (mg N/g soil/day)
DAMOX = rate of change of *Nitrosomonas* (cell/g soil/day)
KDN = decay rate constant of nitrifying organisms (0.05/day)
DNO2OX = rate of change of *Nitrobacter* (cell/g soil/day)

As with other microbial growth processes, the gross growth rates are corrected, for processes other than growth, by a decay factor, KDN for both *Nitrosomonas* and *Nitrobacter*, to calculate the net growth rates, DAMOX and DNO2OX.

Model calculations show an excellent fit with data on nitrification in aqueous systems (Figure 3). However, when using data on the growth rates of the organisms in soil, unexpected accumulation of NO_2 and delay of NO_3 production is predicted. This might be caused by errors in the experimental determination of the values of the specific growth rates in soil.

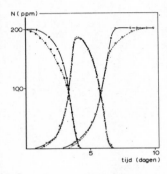

Fig. 3. Comparison of experimental and simulated data on nitrification in aqueous systems.
Solid line: simulation; dotted line: experiment. `..` NH_4^+; x-x = NO_2^-; o-o = NO_3^-.

VOLATILISATION OF AMMONIA

The volatilisation of ammonia is assumed to be controlled by physico-chemical processes, and also that volatilisation only occurs from the surface soil. The concentration of NH_3 in the water-phase is first calculated from the amount of free and exchangeable NH_4^+ in the surface soil. Then the equilibrium between the content of NH_3 in the water-phase and NH_3 in the gas phase is calculated. Instantaneous equilibrium is assumed for this whole sequence with the equilibrium constants, KE and KEG, being dependent on temperature. The dynamic step is the calculation of the amount of NH_3 which diffuses away from the surface soil into the atmosphere.

TABLE 2

EQUATIONS OF THE SUB-MODEL OF VOLATILISATION OF AMMONIA

NH3SOL = CATEXF * (NH4 * CONVFA * OH/KE)/(1 + CATEXF * OH/KE)   (13)

NH3GAS = KEG * NH3SOL   (14)

NH3OUT = DNH3 * NH3GAS * CONVG/DEPTH   (15)

NH3SOL = conc. of NH_3 in soil solution (mg N/ml)

CATEXF = fraction of the total content of ammonium ions in soil which is in solution

CONVFA = conversion factor to convert mg N/g to mg N/ml

OH = OH^--ion conc. in soil (mg/ml)

KE = equilibrium constant for the reaction
$$NH_4^+ + OH^- \rightleftarrows NH_3 + H_2O \quad (16)$$

NH3GAS = NH_3 conc. in gasphase (mg N/ml)

KEG = equilibrium constant for the reaction
$$(NH_3) \text{ solution} \rightleftarrows (NH_3) \text{ gasphase} \quad (17)$$

DNH3 = diffusion rate constant of NH_3 in air (1600 cm^2/day)

NH3OUT = volatilisation rate of gaseous NH_3 (mg N/g soil/day)

CONVG = conversion factor to convert mg N/ml gas/mg N/g)

DEPTH = distance between the place of production of NH_3 and the soil surface (cm)

Figure 4 shows the computed loss of surface applied nitrogen due to ammonia volatilisation depending on temperature. The effect of pH is calculated in Figure 5. These results well agree with those of others (Mills et al., 1974), when losses of NH_3-N were found up to 17% at pH 7.2 and 63% at pH 8.0 - 8.5 during a period of 7 days.

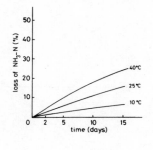
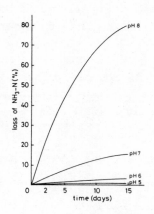

Fig. 4. Computed loss of N due to volatilisation depending on temperature.

Fig. 5. Computed effect of pH on the loss of N due to volatilisation.

FIXATION OF AMMONIUM ON CLAY MINERALS

The fixation of ammonium ions on clay minerals is considered to be a purely physico-chemical process described by the equilibrium equation:

$$NH_4^+ \text{ (free)} \underset{\rightarrow}{\overset{K}{\leftarrow}} NH_4^+ \text{ (fixed)}.$$

This means that the fixation of ammonium is assumed to be a reversible process between fixed NH_4^+ and free NH_4^+, i.e. exchangeable and solute NH_4^+. However, the fixation rate greatly exceeds the rate of release of fixed NH_4^+. Generally, therefore, no dynamic equilibrium exists between fixed and free ammonium ions. The present sub-model is based on a simulation model (Frissel, 1972; Frissel and Reiniger, 1974) of the fixation of potassium ions in soil. Due to the similarity of potassium and ammonium fixation, this seems to be fully justified. The 'fixation' of ammonium ions on soil organic compounds, such as lignin, by chemical processes is not considered.

The calculations were combined with the nitrification sub-model to determine the effect of fixation on nitrification.

Parameter sensitivity analysis showed that only the value of MAXFIX, the maximum content of fixed NH_4^+ significantly affected the nitrification of NH_4^+.

TABLE 3

EQUATIONS OF THE SUB-MODEL OF FIXATION OF AMMONIUM ON CLAY MINERALS

TOFIX = FIXCOF * (NH4 + NH4F)	(19)
TOFIX = MAXFIX , if TOFIX > MAXFIX	(20)
DIFFIX = TOFIX - NH4F	(21)
FRNH4 = DIFFIX * RELEAS , if DIFFIX < 0	(22a)
FRNH4 = 0 , if DIFFIX = 0	(22b)
FRNH4 = DIFFIX * RFIX . if DIFFIX > 0	(22c)

TOFIX = content of NH_4^+ which may be fixed (mg N/g soil)

FIXCOF = coefficient related to equilibrium constant K in equation (18) according to:
FIXCOF = 1/ (K + 1)

NH4F = conc. of fixed NH_4^+ (mg N/g soil)

MAXFIX = maximum conc. of fixed NH_4^+ (mg N/g soil)

FRNH4 = rate of change of fixed NH_4^+ (mg N/g soil/day)

RELEAS = release rate constant (per day)

RFIX = fixation rate constant (per day)

DENITRIFICATION

The term denitrification refers to the microbial use of NO_3^- as a terminal electron acceptor. As this is assumed to occur only under anaerobic conditions, a complete description of denitrification must describe not only the utilisation of NO_3^- under anaerobic conditions, but also the occurrence of these conditions in soil, requiring the behaviour of oxygen to be described. When describing the behaviour of oxygen in soil two cases can be considered: the behaviour of oxygen in completely waterlogged soils and in non-waterlogged soils. In completely waterlogged soils oxygen is supplied only by diffusion through water-filled pores from the atmosphere above the soil. Then transport of oxygen through the soil can be described in a similar way to that of the migration of NO_3^- as used in this model. In non-waterlogged soils the oxygen is supplied via air-filled and water-filled pores. The model considers the diffusion of oxygen from the air-filled pores into the surrounding soil, for which the soil around the air-filled pores, which are considered to be cylindrical, is divided into 10 concentric layers. It is assumed here that the transport of O_2 is only caused by diffusion due to concentration gradients between the layers (equations 43 and 44). The critical calculation in the determination of the occurrence of denitrification is determining the distance between the source of oxygen, viz. the boundary of the air-filled pores, and the point in the surrounding soil where the consumption rate of oxygen exceeds the rate of supply. Beyond that point anaerobiosis will occur and thus denitrification.

The calculation starts with equations (25) - (38) describing the distribution of air-filled pores in the soil. Assuming that the pores are vertical cylinders in the soil, the number of air-filled pores per cm^3, NUMB, is calculated by dividing the total air-filled volume, AFV, by the volume of the pores which are air-filled at any given water content of the soil (equation 40). To calculate this volume the mean radius, MPR, of the air-filled pores at that particular water content is calculated according to equation (24), which virtually gives the minimum radius of pores filled with air at a particular moisture content. The same equation is used (equations 29 and 36) to describe the total pore size distribution in a soil from its pF-curve. The soil volume (of the 1 cm^3 under investigation) is then divided by the number of the air-filled pores (equation 41) and each volume thus obtained, ARAD, is assumed to be around the pores (see Figure 6, which shows the arrangement of 4 pores per cm^3).

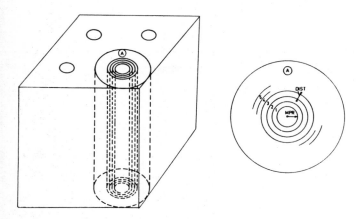

Fig. 6. Schematic picture of the distribution of 4 air-filled pores and the subdividing of the surrounding soil in layers, in a soil volume of 1 cm^3.

If the oxygen concentration, O2C, exceeds a minimum value O_2 will be utilised as terminal electron acceptor (equations 45 and 49); if not, NO_3 will be used by the micro-organisms (equations 45 and 46). The utilisation rates of both O_2 and NO_3, CONSO2 and CONNO3, are proportional to the growth rate, GRB, of the heterotrophic micro-organisms. This growth rate, which is corrected for the oxygen conditions, is calculated in the sub-model for mineralisation and immobilisation based on the availability of carbon. The rate of exchange of O_2 and NO_3 due to transport and consumption are given by equations (51) and (52).

The description given in Table 4 refers to non-waterlogged soils, but the equations for the description of the dynamics of the behaviour of oxygen in soil, as well as of the denitrification, can also be applied with some minor corrections for water saturated soils. The equations shown in Table 4 are the main ones in the sub-model, but do not represent the complete model; therefore the reader is referred to Van Veen (1977).

TABLE 4

EQUATIONS FOR THE SUB-MODEL OF DENITRIFICATION

```
EO2C = PPO2 * H2OC * CONVDF/(32 * HK)                           (23)
MINR = 2 * SIGMA/(10 ** PF * CORF)                              (24)
SUM = 0                                                         (25)
L = 1                                                           (26)
1   Q = L                                                       (27)
    COR(L) = AFGEN(PFT,Q)                                       (28)
    RAD(L) = 2 * SIGMA/(10 ** COR(L) * 1.E-3)                   (29)
    IF(RAD(L).GT.MINR) GO TO 2                                  (30)
    L = L + 1                                                   (31)
    GO TO 1                                                     (32)
2   DO 3 K = L,X                                                (33)
    Z = K                                                       (34)
    COR(K) = AFGEN (PFT,Z)                                      (35)
    RAD(K) = 2 * SIGMA/(10 ** COR(K) * 1.E-3)                   (36)
3   SUM = SUM + RAD(K)                                          (37)
    MPR = SUM/(X-L)                                             (38)
AFV = PORV - WC                                                 (39)
NUMB = AFV/(3.14 * MPR ** 2)                                    (40)
ARAD = SQRT (1/(3.14 * NUMB)                                    (41)
DIST = (ARAD - MPR)/10                                          (42)
DIFO2(I) = DO2 * (O2C(I-1)-O2C(I)) * SURF(I) * BD/DIST          (43)
DIFNO3(I) = DIFN * (NO3(I-1)-NO3(I)) * SURF(I) * BD/DIST        (44)
IF (O2C(I).GE.MINO2) GO TO 30                                   (45)
CONNO3(I) = GRB * YNO3                                          (46)
CONSO2(I) = 0                                                   (47)
GO TO 24                                                        (48)
30  CONSO2(I) = GRB * YO2                                       (49)
    CONNO3(I) = 0                                               (50)
    DO2C(I) = DIFO2(I) - DIFO2(I + 1))/VOL(I)/BD - CONSO2(I)    (51)
    DNO3(I) = (DIFNO3(I) - DIFNO3(I + 1))/VOL(I)/BD - CONNO3(I) (52)
```

EO2C = oxygen concentration in an infinitely small layer around an air-filled pore (mg/g soil)

PPO2 = partial pressure of oxygen in air (0.2)

H2OC = amount grmols H_2O per litre water (55.6 grmol/l)

CONVFD = conversion factor (mg O_2/ml → mg O_2/g soil)

TABLE 4 (CONT)

HK = Henry's Law constant, which depends on temperature

MINR = minimum radius of pores filled with air at a moisture suction PF

SIGMA = surface tension of water (70 dynes/cm)

PF = moisture suction

CORF = correction factor (dynes → mbar)

SUM = sum of pores filled with air

L, Q, COR, K, X and Z are dummy variables

RAD = pore radius (cm)

MPR = mean radius of pores filled with air (cm)

AFV = total air-filled volume (cm^3)

PORV = total pore volume (cm^3)

WC = water content (cm^3)

NUMB = number of air-filled pores in 1 cm^3

ARAD = outer radius of the soil volume around an air-filled pore (cm)

DIST = thickness of one layer around an air-filled pore (cm)

DIFO2 = diffusion rate of oxygen from layer I-1 to layer I (mg O_2/day)

DO2 = diffusion coefficient of oxygen in soil (cm^2/day)

O2C = oxygen concentration (mg O_2/g soil)

SURF = surface through which diffusion occurs (cm^2)

BD = bulk density of soil (g/cm^3)

DIFNO3 = diffusion rate of NO_3^- from layer I-1 to layer I (mg N/day)

DIFN = diffusion coefficient of NO_3^- in soil (cm^2/day)

MINO2 = minimum oxygen concentration for aerobiosis (mg O_2/g soil)

CONNO3 = NO_3^- consumption rate (mg N/g soil/day)

GRB = growth rate of heterotrophic microflora (mg N/g soil/day)

YNO3 = NO_3 utilisation rate constant (5 mg NO_3-N/mg biomass-N)

CONSO2 = O_2 consumption rate (mg O_2/g soil/day)

YO2 = O_2 utilisation rate constant (1.0 mg O_2/mg biomass-N)

DO2C = rate of change of oxygen concentration (mg O_2/g soil/day)

VOL(I) = volume of layer I (cm^3)

DNO3 = rate of change of NO_3^- concentration due to denitrification (mg N/g soil/day)

A typical picture of the O_2 concentrations in several layers around an air-filled pore in a soil with average microbial activity at a water content of 90% saturation is given in Figure 7. Changes of the NO_3^- concentration in this soil with time are shown in Figure 8. Lack of experimental data and problems on the programming aspects hamper a further development and improvement of this sub-model up till now.

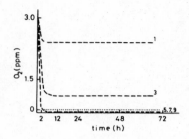

Fig. 7. Simulated O_2 concentrations in the layers 1, 3, 5, 7 and 9 around an air-filled pore at 90% of the water saturation capacity.

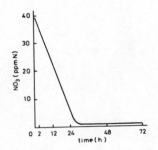

Fig. 8. Simulated NO_3^- concentrations in a 1 cm^3 soil volume at 90% of the water saturation capacity.

MINERALISATION AND IMMOBILISATION

This sub-model, which can be considered to be the core of the total N-model, describes the influence of the C-cycle on the N-transformations in soil. Due to the complexity of the

programme it is beyond the scope of this paper to give detailed
listing of this sub-model.

Based on tests with experimental data, several versions
of this sub-model were subsequently developed. These versions
differed in the way the availability of C-compounds as substrate for micro-organisms was taken into account, but in all
of them the heterotrophic biomass was described as the driving
force behind the mineralisation and immobilisation of C and N.
Carbon and nitrogen are always considered to be the growth
limiting substrates, with N being limiting only at very low
concentrations (1 - 2 ppm). In the first version (Van Veen
and Frissel, 1975) differences in decomposition rate between
C-compounds in natural products such as straw and manure were
accounted for by using an 'availability factor'.

The growth rate of micro-organisms was then

$$\frac{dx}{dt} = \mu^{max} \cdot \frac{C * fa}{Kc + (c * fa)} \cdot x \qquad (53)$$

where x = biomass
 μ^{max} = maximum specific growth rate
 C = C-content
 Kc = saturation constant
 fa = availability factor.

Use of equation (53) for the growth of the microbes, and
thus for the decomposition of organic matter, implies a
sequence in time for the decomposition of the different
compounds, starting with the readily decomposable and finishing
with the less readily decomposable compounds. However, fa does
not really have a well defined physical or biological meaning,
although the authors consider this to be a very important
condition for all parameters and variables of the model
(Frissel and Van Veen, 1978). Therefore, a second set of
descriptions was made in which several compounds are separately
described. The organic matter is assumed here to consist of

proteins, carbohydrates, (hemi)-cellulose and lignin, as shown
schematically in Figure 9 in connection with the other processes
described in the model. The growth of the biomass on these
compounds is described by Michaelis-Menten kinetics as in
equation (53), but use is made of specific parameters for each
compound. The last description is also better from a computer
programming point of view when organic matter is added
continuously, for instance in the case of grazed pastures.

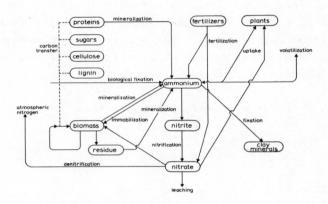

Fig. 9. Scheme of the second version of the mineralisation-immobilisation
sub-model in connection with the other processes of the N-cycle.

Each compound is assumed to be utilised as substrate by
a fraction of the biomass. The size of this fraction is
linearly proportional to the ratio of the content of the
compound degraded by the particular fraction and the total
C-content. Several descriptions of microbial growth using
Michaelis-Menten type of kinetics are proposed in the
literature. All contain serious imperfections. For instance,
in this description the assumption that all microbes involved
have the capacity to use all compounds as substrate certainly
is not correct for cellulose and lignin, but seems reasonable
for more readily decomposable compounds such as carbohydrates.
This could lead to an over-estimation of the rate of

decomposition of cellulose and lignin, but to take account of qualitative differences between microbes would go too much into detail. The present description, however, seems superior for soil microbial processes to the ones in which all substrate is degraded by all microbes at the same place and at the same moment, as proposed for the description of activated sludge processes (Curds, 1974; Stumm-Zollinger, 1966).

In this second version only freshly added organic C was considered. Based on the results of testing this model with experimental data (Figures 11 and 12), this appeared to be erroneous and thus an improved version was developed (Figure 10).

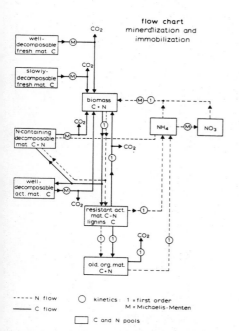

Fig. 10. Scheme of the third version of the sub-model of mineralisation and immobilisation.

This third version differs from the second version only in respect of the more appropriate description of the several forms of C in soils, including freshly added and native soil organic-C. Comparisons with experimental data show this last version to be an improvement compared with the second version, but it is still not a perfect model.

EXPERIMENTAL TEST OF THE MODEL

Experimental verification is a pre-requisite to the development of a model. Therefore, a combined field/greenhouse experiment was carried out mainly to test the second version of the mineralisation and immobilisation sub-model. The following is a summary of the experiment, described by Van Veen (1977).

The soil was a sandy soil (fraction > 50 μm 95.1%) containing 1.23% total-C and 0.06% total N content. It had a CEC of 2.8 meq per 100 g dry soil and pH (KCl) was 4.6. The experimental field was divided into 16 plots, each of 25 m^2. Four treatments (with four replications) were used:

NS : 75 kg N/ha (supplied as $(NH_4)_2SO_4$) + 8 000 kg barley straw/ha
N : 75 kg N/ha (supplied as $(NH_4)_2SO_4$)
S : 8 000 kg barley straw/ha
O : control.

The straw was incorporated by cultivating and the soil was left fallow. Soil samples were taken from ten 10 cm layers every two weeks for a period of 125 days. They were analysed for total N, Cl^- and pH. Biomass was determined only for the layers 0 - 10 cm, 10 - 20 cm, 30 - 40 cm and 70 - 80 cm. Rainfall was recorded daily; air temperature immediately above the soil surface was recorded continuously. The ground water level was determined every 14 days.

The greenhouse experiment was carried out in Mitscherlich pots. The treatments were similar (NS, N, S and O). Measurements of total inorganic nitrogen, NH_4^+ and biomass were carried out at 0, 3, 7, 10, 15, ..., 122 days. Water content was kept at about 60% of the water holding capacity and temperature varied between $22°$ and $25°$ C.

Inorganic nitrogen, NH_4^+, Cl^- and pH were determined by conventional methods. Determination of the biomass was based on direct microscopic counting after staining with fluorescein-isothiocyanate (FITC) (Babiuk and Paul, 1971). For simulating the sub-models for nitrification, mineralisation, immobilisation and transport as well as that of volatilisation of ammonia, only results from the greenhouse experiments were used. Comparison of the experimental data on inorganic nitrogen with the simulated ones (Figures 11a and b, 12a and b) for both the greenhouse and the field (0 - 10 cm) experiment shows rather serious disagreements, which also showed up when comparing experimentally determined and simulated biomass vs time.

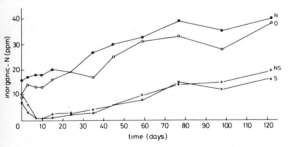

Fig. 11a. Experimentally determined inorganic N-conc. vs time in a greenhouse experiment at constant temperature and moisture content. NS = nitrogen + straw; N = nitrogen; S = straw; O = control.

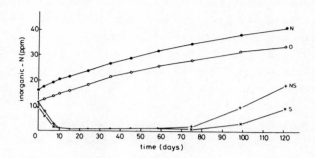

Fig. 11b. Simulated inorganic N-conc. vs time in a greenhouse experiment. (See Figure 11a).

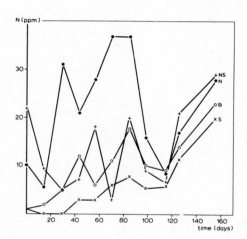

Fig. 12a. Experimentally determined inorganic N-conc. vs time in a field experiment. (See Figure 11a).

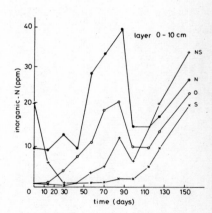

Fig. 12b. Simulated inorganic N-conc. vs time in a field experiment. (See Figure 11a).

With the third version of the sub-model of mineralisation and immobilisation, Figure 13 shows that a better fit was obtained for the inorganic nitrogen behaviour in the NS treatment of the greenhouse experiment.

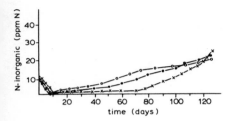

Fig. 13. Comparison of simulated and experimentally determined inorganic N-conc. in soil vs time after application of straw and N

  o - o   experiment
  x - x   simulation with second version of the model
  ● - ●   simulation with third version of the model

Further testing of this model with data from fallow, recently reclaimed soils of the Lake IJssel polders obtained from the Rijksdienst voor de IJsselmeerpolders, Lelystad, shows a satisfactory fit (Figure 14) but some imperfections remain.

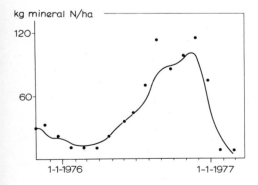

Fig. 14. NO_3 conc. vs time for a fallow plot in the IJsselmeerpolders.
  .... = experiment; _____ = simulation.

Further development of the model will involve more experimental verification including plant growth as well as theoretical development of sub-models on the effect of plants on the soil N-transformations and on N_2 fixation. Moreover, laboratory and field experiments will be carried out to obtain more insight into the value of data from laboratory experiments on microbial growth and to obtain new appropriate data on microbial turnover, which appears a major parameter in this model and generally in the role of micro-organisms in nutrient cycles in the soil.

REFERENCES

Babiuk, L.A. and Paul, E.A., 1970. The use of fluorescein-isothiocyanate in the determination of the bacterial biomass of grassland soil. Can. J. Microbiol. 16: 57-62.

Beek, J. and Frissel, M.J., 1973. Simulation of nitrogen behaviour in soils. Pudoc, Wageningen, p. 67.

Curds, C.R., 1974. Computer simulations of some complex microbial food chains. Water Res. 8: 769-780.

Frissel, M.J., 1972. Model calculations on the vertical transport of potassium ions in soil. Proc. 9th Colloq. Int. Potash Institute. Landshut, F.R.G., p. 157-170.

Frissel, M.J. and Reiniger, P., 1974. Simulation of accumulation and leaching in soils. Pudoc, Wageningen, p. 116.

Frissel, M.J. and van Veen, J.A., 1978. A critique of computer simulation modelling for nitrogen in irrigated croplands. Computer simulation of nitrogen behaviour in soil. In: 'Nitrogen in the environment'. Vol. 1. (Eds: D. Nielsen and J.G. MacDonald). Acad. Press, Inc. New York, p. 145-162.

Frissel, M.J., van Veen, J.A. and Kolenbrander, G.J., 1976. The use of sub-models in the simulation of nitrogen transformations in soils. Presented at the I.S.S.S. Congress on 'Agrochemicals in soils', Jerusalem.

Mills, H.A., Barker, A.V. and Maynard, D.N., 1974. Ammonia volatilisation from soils. Agron. J. 66: 355-358.

Stumm-Zollinger, E., 1966. Effects of inhibition and repression on the utilisation of substrates by heterogeneous bacterial communities. Appl. Microbiol. 14: 654-664.

Tanji, K.K. and Gupta, S.G., 1978. Computer simulation modelling for nitrogen in irrigated croplands. In: 'Nitrogen in the environment'. Vol. 1. (Eds: D. Nielsen and J.G. MacDonald). Acad. Press, Inc. New York.

van Veen, J.A., 1977. The behaviour of nitrogen in soils. A computer simulation model. Ph. D. Thesis, Free Reformed University, Amsterdam.

van Veen, J.A. and Frissel, M.J., 1975. Computer simulation for the behaviour of nitrogen in soil and leaching to groundwater. Ext. Report No. 30. Association Euratom-ITAL, p. 31.

NITRATE PRODUCTION, MOVEMENT AND LOSSES IN SOILS

J.K.R. Gasser
Agricultural Research Council, London, UK.

Slurry consists of faeces and urine together with varying amounts of other organic materials, such as bedding and spilt feed, and water. The amount of water added will vary from zero or very little to several times the volume of slurry depending on the conditions. Nitrogen in the faeces and urine is almost entirely in organic forms, with a little ammonium. Oxidised forms are normally absent. Urea nitrogen predominates in the urine and more complex forms in faeces (Doak, 1952; Petersen et al., 1956). However, the urea is decomposed rapidly together with some of the other readily degraded organic compounds to yield ammonia so that urea is effectively absent from slurry as handled. These processes together with the following ones are represented schematically in Figure 1 and we may consider the information available to allow the model to be developed quantitatively.

Urea is decomposed rapidly and ammonium carbonate formed increases the pH of the slurry and there is usually some loss of nitrogen as ammonia. The amount lost depends on the pH, surface area and amount of aeration. The formation of nitrate during the storage or treatment of slurry depends on the treatment and was extensively discussed at the recent EEC Seminar on, 'Engineering problems with effluents from livestock', 17 - 21 September 1978, Cambridge, England. This aspect will not be discussed here.

EFFECTS OF pH

The losses of ammonia from slurries applied to soils has not been measured but useful analogues may be drawn from the loss of ammonia from urea supplied to soils which has been extensively studied. Two important factors were found to be the cation exchange capacity of the soil and drying of the soil after applying urea to the surface. Aeration of an alkaline solution of ammonium salts will also increase losses of ammonia so that wind will also increase losses from surface applied slurry and these losses will decrease the total amount of ammonium available for nitrification and decrease the pH of the slurry. In the systems investigated using pure urea, nitrite tended to accumulate at alkaline pHs, whereas nitrification is most rapid and proceeds readily to nitrate at or near pH 7. Nitrite is undesirable both because it is toxic to plants and because it is at risk of loss by denitrification.

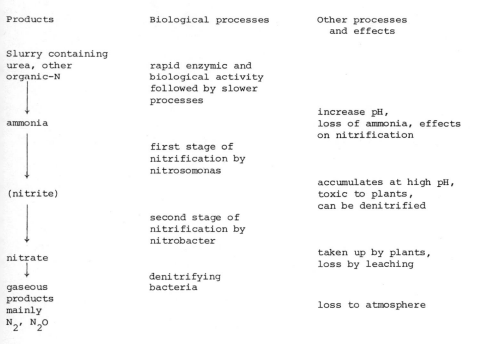

Fig. 1. A model scheme for the production, movement and loss of nitrate from animal slurries.

MOVEMENT AND DECOMPOSITION OF SLURRY APPLIED TO SOILS

Slurry may be applied to the surface of the soil or injected into it. Because of its semi-liquid form problems associated with slurry application are accentuated. For example when slurry is applied to grassland, organic material accumulates on the surface and is encouraged because aerobic decomposition of slurry solids is delayed until considerable drying has taken place (McAllister, 1976). Table 1 gives results also from McAllister (1976) showing that organic matter accumulates in the surface layers of soil following repeated additions of pig slurry to grassland. The passage of organic matter and nutrients, including nitrate, into drainage water has been shown (McAllister, 1976; O'Callaghan and Pollock, 1976); even though slurry was applied in accordance with a landspreading model which only permitted slurry to be applied when there was a deficit of soil moisture (O'Callaghan and Pollock, 1976). Table 2 gives values from McAllister (1976) and shows how drain depth affects the amount of slurry entering the drainage water. The deep drains were at 1.25 m and back filled with soil and the shallow drains were at 0.5 m and filled to near the surface with coarse stones. When the slurry was applied there were steady and similar flows of about 1 000 l/h at both sites. Effects on the deep drainage water were slight, whereas there were marked effects on water from the shallow drains. There was evidence from the COD values and P content that solid matter was entering the shallow drainage before the application of slurry, although values measured by O'Callaghan and Pollock (1976) showed that less than two per cent of the activity of the slurry was found in the drainage water.

Burford et al. (1976) applied cattle slurry at rates up to 500 t/ha. The slurry had 15.6% DM containing 2.15% N. The maximum rate therefore supplied 1 840 kgN/ha. This may be compared with the 850 kgN/ha in dung patches and 450 kgN/ha in urine patches of cattle (Petersen et al., 1956). Slurry is a mixture of urine and faeces, and the maximum rate used by Burford et al. (1976) is somewhat greater than the simultaneous voiding of urine and faeces at one point. Although excessive as a continuous dressing, the results provide useful information to develop a model to describe effects in the slurry layer and in the soil underneath as discussed by the authors. Nitrate

TABLE 1

TOTAL CARBON AND NITROGEN CONTENTS OF A GRASSLAND SOIL WITHOUT SLURRY AND WITH 212 m^3/ha OF SLURRY ANNUALLY FOR EIGHT YEARS (McAllister, 1976)

Soil depth mm	Carbon in soil		Nitrogen in soil	
	Control %	with slurry %	Control %	with slurry %
0 - 150	4.47	8.08	0.58	0.71
150 - 300	2.03	2.16	0.21	0.25
300 - 460	0.77	0.87	0.07	0.11
460 - 610	0.36	0.46	0.03	0.04
610 - 760	0.27	0.37	0.02	0.03

TABLE 2

COMPOSITION OF DRAINAGE WATER FROM DEEP AND SHALLOW DRAINS BEFORE AND AFTER APPLICATION OF PIG SLURRY (McAllister, 1976)

Hours after spreading	BOD mg/l		N mg/l	
	Drain depth			
	Deep	Shallow	Deep	Shallow
Nil	3	4	18	6
0.5	10	530	18	91
1	5	215	19	36
2	3	34	17	12
4	2	7	8	6
8	2	7	10	7
24	neg	3	10	5
48	neg	3	7	5

initially present in the soil and slurry was lost by denitrification, and nitrate accumulation was inhibited for some time by either poor aeration *per se* or products resulting from restricted aeration. Gradual improvement of aeration, with time, from the slurry surface downwards presumably allowed nitrification to proceed and nitrate to accumulate through an increasing depth of soil. However, the presence of a zone of poorer aeration and intense reducing activity below the zone of nitrate accumulation, and the occurrence of N_2O in both these zones following rainfall, indicate that denitrification lessened the amount of nitrate that was leached for a period after the heavy application.

These ideas are illustrated in Figure 2, where the dotted area represents the reducing zone preventing nitrate movement downwards and the shaded area represents the zone of intermittent or slow gaseous N loss depending on soil moisture conditions.

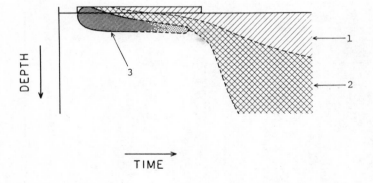

Fig. 2. Diagram illustrating the pattern of nitrate accumulation and loss following a heavy slurry application (Burford et al., 1976)

1 = Nitrate accumulates - in surface soil due to nitrification. Nitrate decreases due to leaching; <u>slight</u> losses at N_2O by denitrification at high moisture content.
2 = Nitrate accumulates - in subsoil due to leaching; <u>larger</u> denitrification losses.
3 = Zone of intense reducing activity. Nitrate does not accumulate; leached nitrate is denitrified.

The results of Burford et al. (1976) showed that there were appreciable losses of N following a heavy application of slurry to a light-textured soil. Nitrate found in the drainage water accounted for less than one per cent of the total-N applied. Detailed studies are needed to establish losses in percolating water not intercepted by drains.

The soil atmosphere under the slurry contained up to 680 ppm (v/v) of N_2O. Gaseous loss of nitrogen was probably significant, but further work needs to be done on gas transfer processes before the magnitude can be assessed. For inorganic nitrate applied in late spring, concentrations of nitrous oxide indicated only minor losses by denitrification.

The medium term decomposition of slurry applied to soils has not been studied, but, as already noted in Table 1, the continued addition of pig slurry to a grassland soil for eight years increased the percentage organic matter and nitrogen particularly in the surface 150 mm (McAllister, 1976). Before reliable models can be constructed, experimental results over many years work are required and this requires considerable resources. For example, in a study of the decomposition of plant material in soils over 10 years, Figure 3 from Jenkinson's (1977) results shows that the decay was explained better by a double exponential model than by a single exponential. The double exponential had the form:

$$y = (70.9 \pm 0.8) \exp[-(2.83 \pm 0.15)t]$$
$$+ (29.1 \pm 0.7) \exp[-(0.087 \pm 0.004)t]$$

where y = percent labelled C remaining and t = time in years. This model predicts that about 70% of the plant material decomposes with a half-life of 0.25 years and the remainder with a half-life of 8 years. This is an over simplification of the real processes of decay in soil because it takes no account of the formation and decay either of the biomass or of very inert organic matter.

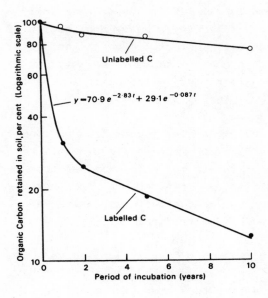

Fig. 3. Losses of labelled and unlabelled C from Rothamsted soils; note logarithmic scale. (Jenkinson, 1977).

For slurry injected into soils, including under grassland, information is lacking on its distribution in the soil and mode of decomposition. In a study on the fate of aqueous ammonia injected under grassland, Gasser and Ross (1973) described an experimental method for taking a cross-section of the soil around the point of injection and a mathematical model for calculating lines of equal concentration. They also found evidence for the movement of ammonia solution through structural voids in the soil and deflocculation of clay by the alkaline solution of ammonia. Many slurries are equivalent to dilute solutions of ammonia containing organic matter and similar methods could be used to investigate their decomposition in soil and effects on soil processes.

These results indicate the need to develop models quantifying the decomposition of slurry on and in soils, its movement through structured soils and the effects of the resultant organic matter and nitrogen both on the concentrations of nitrate in the lower soil layers and soil conditions leading to denitrification.

NITRIFICATION AND LEACHING

The processes of nitrification of ammonium in soil have been extensively studied as has been the leaching of nitrate in soil and its effects on the availability of nitrate to crops. Burns (1977) recently reviewed some of these areas stressing that nitrification follows a pseudo-first order rate equation which was confirmed for nitrification of ammonium in a sandy-loam soil. Integration of the equation gives:

$$\ln So - \ln S = Kt$$

where So is the inital concentration of substrate (ammonium), S is the concentration at time t and K is the apparent rate constant. The value of K varies with temperature and Page (1975) extrapolated the results of Tyler et al. (1959) to derive the empirical equation:

$$K = Ko \times 2.1^{0.1T}$$

where Ko is the value of K at $0^{o}C$ and T is the temperature in ^{o}C (which must be in the range of normal soil temperatures). These equations can be used to give approximate estimates of the loss of ammonium ions from the soil by nitrification, provided the basic assumptions of the model hold and soil moisture and pH do not fluctuate greatly.

For the leaching of nitrate from soils, Burns (1977) proposed a simple mathematical model which assumes that nitrate movement from a layer of soil only occurs when the field capacity is exceeded and is direcly proportional to water movement. In its simplest form the model can be expressed as single equations which enable leaching predictions to be made for different initial distributions of nitrate in the field. Provided the soils are of fairly uniform field capacity with depth and the initial water content of the soil is close to field capacity, these equations may all be rearranged to the following approximate form:

$$f \simeq \left(\frac{100P}{100P + V_m}\right)^x$$

where f is the fraction of nitrate leached below a depth h cm in a soil of field capacity V_m (% by volume) after P cm of drainage water has passed through the profile and where x is a simple function of h. The nature of x depends on the initial distribution of nitrate in the profile; $x = h$ for surface-applied material, $x = h - w/2$ for nitrate incorporated uniformly to a depth w cm in the soil and $x = h/2$ for uniformly distributed nutrient. These initial distributions of nitrate should cover most of the commonly encountered nitrate distributions in the field. Values of P in the above equation may be estimated from a simple water balance equation:

$$P = R - E - Do$$

where R and E are the cumulative rainfall and evapotranspiration and Do is the initial water deficit of the soil. The validity of the leaching equation was tested using experimental results on soils ranging from a sand to a clay, where good agreement was found between the model and experimental data.

In another model of nitrate movement in soils, Wild and Babiker (1977) found that on bare fallow plots given high rates of nitrate, the depth of the peak concentration of nitrate was correlated with the apparent depth of movement of the winter rainfall calculated as Q/θ, Q being the amount of effective rainfall (total rainfall - evaporation), and θ the volumetric water content of the soil. The regression equation was:

$$y = 0.57x + 4.1$$

where y is the depth of the nitrate peak (cm) and x is Q/θ (cm); that is, the depth of movement of nitrate was about 0.6 of the apparent depth of movement of the rainfall which followed the application of the fertiliser. The recovery of nitrate in the top 50 cm of the soil profile at the end of winter also varied with Q/θ, although no movement apparently occurred until there had been a minimum of 20 cm through drainage.

The influence of organic matter present in slurry on the nitrification of ammonium, the risk of denitrification and the movement of water through the soil affecting the movement of nitrate are not known. There is evidence that alkaline solutions of ammonia can deflocculate clay in clay soils (Gasser and Ross, 1973), and the presence of finer slurry solids blocks soil pores (McAllister, 1976) and may influence both nitrification and denitrification (Burford et al., 1976). The injection of slurries under grassland will require further work to elucidate these points. Surface application leads to the formation of a simpler planar model for which some information is already available.

CONCLUSIONS

The work reviewed shows that there is very limited information available on the behaviour and decomposition of slurries applied to soils. Further progress in understanding the processes will require much experimental work, lasting for many years.

The additional information should allow more exact models of slurry performance to be constructed leading to more efficient use of the plant nutrients in the slurry for crop production with decreased risk of polluting surface run-off, or drainage in deep waters.

REFERENCES

Burford, J.R., Greenland, D.J. and Pain B.F. 1976. Effects of heavy dressings of slurry and inorganic fertilisers applied to grassland on the composition of drainage waters and the soil atmosphere. In: Agriculture and Water Quality. Technical Bulletin Ministry of Agriculture, Fisheries and Food No. 32 pp 432-443, H.M.S.O.

Burns, I.G. 1977. Nitrate movement in soil and its agricultural significance. Outlook on Agriculture 9, No. 3, 144-148.

Doak, B.W. 1952. Some chemical changes in the nitrogenous constituents of urine when voided on pasture. Journal of Agricultural Science, Cambridge, 42, 162-171.

Gasser, J.K.R. and Ross, G.J.S. 1973. The distribution in the soil of aqueous ammonia injected under grass. Journal of the Science of Food and Agriculture, 26, 719-729.

Jenkinson, D.S. 1977. Studies on the decomposition of plant material in soil V. The effects of plant cover and soil type on the loss of carbon from ^{14}C labelled ryegrass decomposing under field conditions. Journal of Soil Science, 28, 424-434.

McAllister, J.S.V. 1976. Studies in Northern Ireland on problems related to the disposal of slurry. In: Agriculture and Water Quality. Technical Bulletin Ministry of Agriculture, Fisheries and Food No. 32 pp 418-431. H.M.S.O.

O'Callaghan, J.R. and Pollock, K.A. 1976. A land spreading trial with pig slurry. In: Agriculture and Water Quality. Technical Bulletin Ministry of Agriculture, Fisheries and Food No.32 pp 371-399, H.M.S.O.

Page, E.R. 1975. The location and persistence of ammonia (aqueous, anhydrous and anhydrous + N. Serve) injected into a sandy loam soil as shown by changes in concentrations of ammonium and nitrate ions. Journal of Agricultural Science, Cambridge, 85, 65-74.

Petersen, R.G., Woodhouse, W.W. Jr. and Lucas H.L. 1956. The distribution of excreta by freely grazing cattle and its effect on pasture fertility II, Effect of returned excreta on the residual concentration of some fertilizer elements. Agronomy Journal, 48, 444-449.

Tyler, K.B., Broadbent, F.E. and Hill, G.N. 1959. Low temperature effects on nitrification in four California soils. Soil Science, 87, 123-129.

Wild, A. and Babiker, I.A. 1977. Winter leaching of nitrate at sites in southern England. In: Agriculture and Water Quality. Technical Bulletin Ministry of Agriculture, Fisheries and Food No.32 pp 153-162, H.M.S.O.

DISCUSSION

H. Vetter *(West Germany)*

Dr. van Veen, you said that nitrogen moves horizontally. How much moves and how far because in our field experiments nitrogen increased on the unmanured plots after manuring the neighbouring plots?

J.A. van Veen *(The Netherlands)*

I believe I mentioned the horizontal movement of nitrogen only for denitrification, when it is necessary in order to describe the occurrence of anaerobic zones in the soil. You have to consider horizontal transport because when there are some anaerobic volumes both oxygen and nitrate will move horizontally towards this place because of the concentration gradients caused by their consumption. The Institute for Biological Studies in Wageningen has found an increase of about 100 kg/ha of nitrogen per year in a small plot surrounded by plots 2 or 3 metres away with at least 200 kg nitrogen/ha annually. Whether this nitrogen comes from diffusion or volatilisation of ammonia is not known.

A. Dam Kofoed *(Denmark)*

The reports on the modelling work presented here have been very interesting and I would ask whether these models can be adapted at some time in the future for use in a practical way at farm level. To my mind this should be the objective. The ultimate aim of these models is to benefit the farmer.

NITROGEN MINERALISATION AND NITRIFICATION OF PIG SLURRY ADDED TO SOIL IN LABORATORY CONDITIONS

J.C. Germon*, J.J. Giraud*, R. Chaussod* and C. Duthion**
* Laboratoire de Microbiologie des Sols
** Laboratoire d'Agronomie
INRA, 17, rue Sully, 21034 Dijon Cedex, France.

The fertilising value of pig slurry essentially depends on its nitrogen supplying power as ammonia which may be lost by volatilisation or temporarily immobilised in the soil as microbial biomass. Ammonia is nitrified ultimately to nitrate, which may be assimilated by plants or microflora, or leached, or denitrified. These different pathways are controlled by the environmental factors, especially temperature, water-content and physicochemical properties of the soil.

We did laboratory experiments to study the effects of these different parameters and give some preliminary results on nitrogen mineralisation and nitrification of pig manure applied at different rates to soil incubated at three temperatures.

MATERIALS AND METHODS

The pig slurry used came from fattening pigs. Before use, it was sieved through a 5 mm sieve to remove the straw, and homogenised with a 'Turmix' grinder: its composition is given in Table 1.

Table 2 gives the properties of the soil, as it was used after air-drying and sieving (2 mm).

Incubations were carried out in 2.5 l air-tight boxes, each containing 2 kg of soil. After applying the manure, sufficient water was added to bring the soil moisture content to field capacity. The slurry supplied 0, 127, 381 and 635 mg N (Kjeldahl method) per kg of soil on an air-dried weight soil basis, corresponding to 33, 100, 166 m^3/ha based on the area of the box (186 cm^2). The manure was well mixed with the soil.

The incubations were prepared in triplicate and incubated at 8°C, 20°C and 28°C. The boxes were aerated with an airflow rate of 70 ml/min across the soil surface. The carbon dioxide and ammonia in the air were trapped in sodium hydroxide and boric acid before and after passing through the incubation boxes.

TABLE 1

MANURE COMPOSITION AND ADDED AMOUNTS AT 3 RATES

Determinations	Raw manure concentrations (g/l)	Added amounts (mg/kg of dry soil)		
		rate 1	rate 3	rate 5
Dry matter	34.4	1 143	3 430	5 717
Suspended matter	24.3	810	2 430	4 050
COD	33.1	1 103	3 310	5 517
Kjeldahl-N	3.81	127	381	635
NH_4 - N	2.75	92	275	458
Total P	0.93	31	93	155
Ash	9.85	328	985	1 642
Ca	1.62	54	162	270
Mg	0.44	15	44	73
K	1.98	66	198	330
Na	0.30	10	30	50
Cu	0.027	0.9	2.7	4.5

Duplicate soil samples were taken from each box after 1, 2, 4, 6 and 8 weeks. After extracting with potassium chloride solution, the nitrate and nitrite nitrogen in the extracts were estimated using an autoanalyser (colorimetric method, using diazotisation and complex formation).

The trapped ammonia was estimated with the same apparatus using indophenol complex formation. The ammonium extracted from the soil was analysed by distillation or with the

autoanalyser. The result obtained by the second method was
corrected by a coefficient obtained by comparing the two methods
This provided six elemental analyses for each treatment.

TABLE 2
ANALYSES OF SOIL

Determinations	
Texture %	
0 - 2µ	43
2 - 20µ	31.5
20 - 50µ	20.5
50 - 200µ	3.5
0.2 - 2 mm	1.5
Total carbonate %	<0.6
pH (water)	6.6
C (Ann)%	1.79
N (Kjeldahl) %	0.2
C : N ratio	8.9
P_2O_5 (Joret-Hebert) %	0.003
CEC meq/100 g	26.4
Ca "	27.7
Mg "	1.34
K "	0.30
Na "	0.06
Equiv. Humid. (1 000 g) %	27.9

RESULTS AND DISCUSSION

a) Nitrogen losses by ammonia volatilisation

Under our experimental conditions, results in Table 3
show that negligible amounts of ammonia were lost by volatilisation, although such losses are frequently mentioned in the
literature (Stephens and Mill, 1973; Lauer et al., 1976). The
lack of ammonia volatilisation may be due to mixing the slurry
with soil and maintaining the soil very moist.

TABLE 3

CUMULATIVE AMOUNTS OF AMMONIA-N (mg/kg of dry soil) VOLATILISED DURING 8 WEEKS INCUBATION

Temperature	Ammonium-N, mg/kg, applied to soil in slurry			
	0	98	281	464
	Ammonia-N lost, mg/kg			
$8°C$	0.1	0.1	0.1	0.1
$20°C$	0.2	0.3	0.4	0.4
$28°C$	0.3	0.4	0.6	0.5

b) Total mineral nitrogen in soil

Table 4 gives the ammonium, nitrite, nitrate and total-N contents of the soils, with and without slurry, at the start of the experiment and the various times of sampling, when incubated at $8°C$, $20°C$ and $28°C$. Figure 1 shows the total mineral-N.

At $8°C$ and $20°C$, some soil nitrogen was mineralised during the first week but no more thereafter, whereas at $28°C$, there was further mineralisation during the later part of the incubation. With added slurry, ammonium-N was increased; later results were slightly irregular and there was no evidence for mineralisation of organic-N in the slurry.

Nitrite accumulated during incubation at all temperatures but at the end of the experiment at $20°C$ and $28°C$, the mineral nitrogen is almost entirely as nitrate; at $8°C$, both ammonium and nitrite were also present. The uniformity of the results suggests that little nitrate or nitrite was lost by denitrification. Other workers under closely related conditions, observed nitrogen immobilisation phases, which were sometimes very important (Maass et al., 1973; Sowden, 1976). This is not the case here and our results are nearer to those observed by Cooper (1975). The origin of immobilisation and turnover of N will have to be studied, particularly because of its agricultural importance.

TABLE 4

AMOUNTS OF AMMONIUM, NITRITE, NITRATE AND TOTAL MINERAL NITROGEN, mg/kg OF DRY SOIL, IN SOIL WITH AND WITHOUT ADDED PIG SLURRY INCUBATED FOR VARIOUS PERIODS AT THREE TEMPERATURES

Period of incubation	Form of mineral nitrogen	8°C				20°C				28°C			
		\multicolumn{12}{c}{Equivalent amount of slurry added m³/ha}
		0	33	100	166	0	33	100	166	0	33	100	166
0	NH_4	7	98	281	464	7	98	281	464	7	98	281	464
	NO_2	0	0	0	0	0	0	0	0	0	0	0	0
	NO_3	80	80	80	80	80	80	80	80	80	80	80	80
	Total	87	179	361	544	87	178	361	544	87	178	361	544
1	NH_4	24	111	377	403	14	67	262	452	19	46	262	409
	NO_2	8	7	4	3	10	33	30	24	1	26	42	48
	NO_3	80	83	93	92	97	101	90	91	94	115	102	86
	Total	112	201	374	498	121	201	382	567	114	187	406	543
2	NH_4	2	72	303	441	3	23	146	373	1	ND	121	341
	NO_2	12	15	9	6	0	52	63	66	0	0	36	66
	NO_3	89	90	96	95	108	133	126	113	109	190	206	178
	Total	102	177	408	542	111	208	335	552	110	ND	363	585

(Table 4 Cont.)

TABLE 4 (Cont.)

Period of incubation	Form of mineral nitrogen	Temperature of incubation											
		8°C				20°C				28°C			
		\multicolumn{12}{c	}{Equivalent amount of slurry added m³/ha}
		0	33	100	166	0	33	100	166	0	33	100	166
4	NH$_4$	3	52	ND	ND	3	3	23	150	4	3	3	5
	NO$_2$	11	39	35	29	0	0	12	40	0	0	0	3
	NO$_3$	86	79	81	87	104	211	322	269	130	220	404	596
	Total	100	170	ND	ND	107	214	357	459	134	223	407	604
6	NH$_4$	1	31	ND	ND	1	0	1	1	1	1	1	1
	NO$_2$	9	61	70	64	0	0	0	0	0	0	0	0
	NO$_3$	94	83	91	87	114	211	372	574	125	237	463	622
	Total	104	175	ND	ND	115	211	373	575	126	238	464	623
8	NH$_4$	1	11	ND	ND	0	1	1	2	4	4	1	1
	NO$_2$	5	75	104	104	0	0	0	0	0	0	0	0
	NO$_3$	98	111	114	116	120	217	385	595	140	223	394	609
	Total	104	197	ND	ND	120	218	386	597	144	227	395	610

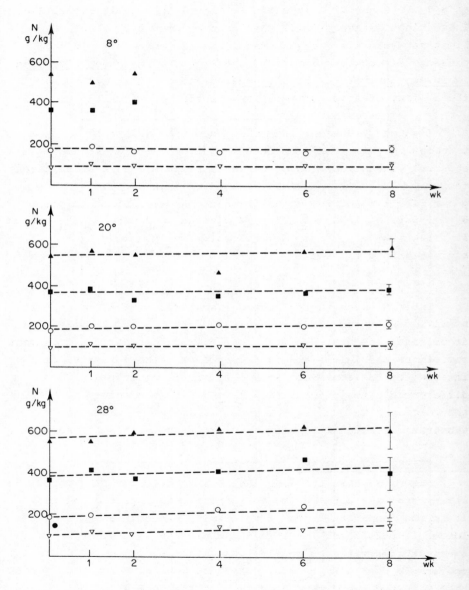

Fig. 1. Total mineral nitrogen formation: rates: 0 ▽, 33 ○, 100 ■, 166 m³/ha

c) Nitrification

Kinetics of nitrification

The production of nitrites and nitrates during incubation at $8°C$, $20°C$ and $28°C$ is shown in Figures 2, 3 and 4 respectively. These kinetics approximate to microbial growth in which the limiting factor is the substrate, that is ammonium. In the presence of adequate ammonium, the <u>amount of nitrogen nitrified is independent of the amount applied</u> and probably depends on the ability of ammonium-oxidising microflora to multiply.

Figures 2, 3 and 4 show the pronounced effect of temperature which could be described by the change in the apparent generation time of the nitrifying bacteria at different temperatures. Our experimental results suggest that the generation time of the ammonium-oxidising microflora is about 8 to 9 days at $28°C$ and 10 to 14 days at $20°C$ with very short, if any, lag phases. At $8°C$, the results do not allow such approximations to be made - they require the incubations for a longer time.

Table 5 gives the statistical analyses of the amounts of nitrified nitrogen for the results obtained after 2 weeks incubation, which agrees with the interpretation. The treatment variations are mainly due to temperature. There is also an interaction between amounts applied and temperatures. No differences between the three amounts can be observed at $8°C$ and $20°C$. At $28°C$, only the smallest amount in which the substrate is completely used, differs from the two others.

d) Nitrite accumulation

Figure 5 shows the accumulations of nitrite. The more slurry that was applied the more nitrite accumulated and the lower the temperature, the more nitrite was formed and the longer it persisted. So that least nitrite was found for the shortest time with the smallest amount of slurry at $28°C$ and conversely most nitrite was formed and persisted for the longest time with the two largest amounts of slurry added to soil incubated at $8°C$.

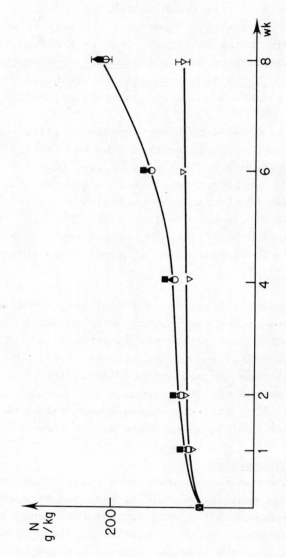

Fig. 2. Nitrogen ($NO_2 + NO_3$) production at $8°C$.

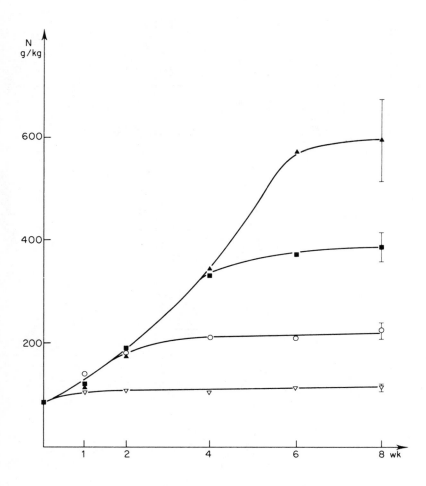

Fig. 3. Nitrogen (NO_2 + NO_3) production at $20°C$.

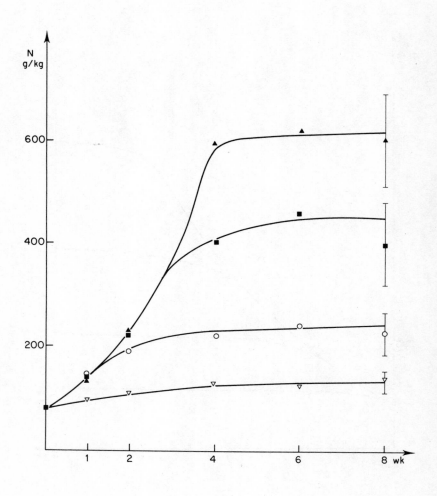

Fig. 4. Nitrogen ($NO_2 + NO_3$) production at $28°C$.

TABLE 5

ANALYSIS OF VARIANCE OF THE AMOUNTS OF NITROGEN NITRIFIED AFTER 2 WEEKS INCUBATION

Source	Sums of squares	Degrees of freedom	Mean squares	Calculated F value
Treatments				
Temperatures	139 663.5	2	69 831.7	325.8**
Rates applied	4 075.0	2	2 037.5	9.5**
Interaction (Temp. x rates)	8 901.7	4	2 225.4	10.4**
Residual	9 645.2	45	214.3	
Total	162 285.4	53		

These observations agree with other results from the literature. The nitrite-oxidising microflora (*Nitrobacter* spp) are more sensitive to temperature decrease than the ammonium-oxidising microflora (*Nitrosomonas* spp). Moreover, Nitrobacter is inhibited by large amounts of ammonium-N in soils (Dommergues, 1970).

On the other hand, our results are the reverse of those expected from the model of Laudelout et al. (1976), which indicates that nitrite accumulates only at temperatures above $20^\circ C$. This shows that insufficient is known about all the parameters needed to build up a valid model under very different conditions.

e) Nitrification balance

Table 6 gives the amounts of nitrate + nitrite nitrogen after 2 months incubation from the added slurry and the percentages of the total and ammonium nitrogen that were nitrified. In the soils with complete nitrification, the nitrified nitrogen accounts for from 65 to 77% of the total nitrogen and 90 to 104% of the ammonium nitrogen in the manure.

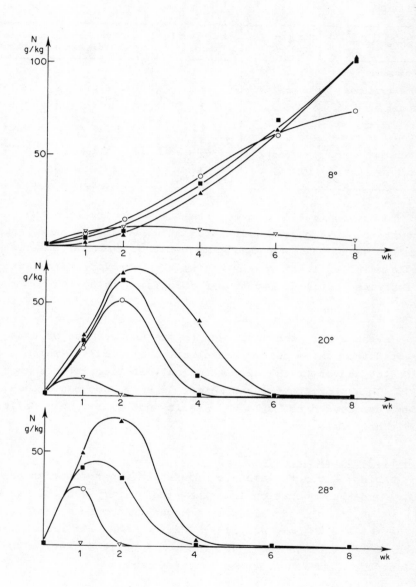

Fig. 5. Nitrite production at 3 temperatures.

TABLE 6

AMOUNTS OF NITROGEN ADDED AS SLURRY NITRIFIED AFTER 8 WEEKS INCUBATION

T°C	Slurry applied equivalent to m³/ha	NO_2-N + NO_3-N formed after 8 weeks (mg/kg of dry soil)	Nitrified-N as a percentage of added	
			Kjeldahl-N %	Added NH_4^+-N %
8°C	0			
	33	93	73	101
	100	115	30	41
	166	117	18	26
20°C	0			
	33	98	77	107
	100	266	70	97
	166	476	75	104
28°C	0			
	33	83	65	90
	100	254	67	92
	166	468	74	102

CONCLUSIONS

From our experiments, we reach the following conclusions:
- When slurry is mixed with soil very little ammonia is lost by volatilisation.
- The mineralisation of organic nitrogen in slurry is slow because the total amount of mineral nitrogen varied little during 8 weeks incubation.
- The nitrification process follows the kinetics of microbial growth. The apparent generation time for the ammonium-oxidising microflora is between 8 and 9 days at 28°C and 10 and 14 days at 20°C.
- Nitrite accumulates during nitrification; the amount increasing with the amount of slurry applied and the persistence of nitrite increases as the temperature decreases.

REFERENCES

Cooper, J.E. 1975. Soil Biol. Biochem. $\underline{7}$, 119-124.

Dommergues, Y. and Mangenot, F. 1970. Ecologie microbienne du sol, Masson, Paris.

Laudelout, M., Lambert, R. and Pham, M.L. 1977. In 'Utilisation of manure by landspreading'. Commission of the European Communities. Co-ordination of agricultural research. Eur. 5672e., pp 447-457.

Lauer, D.A., Bouldin, D.R. and Klausner, S.D. 1976. Journ. Env. Qual., $\underline{5}$, 2, 134-141.

Maass, G., Kaunat, M. and Langecker, W. 1973. Arch. Acker-und-Pflanz. Bodenk, $\underline{17}$, 7-8, 649-659.

Sowden, F.J. 1976. Can. Journ. Soil Sci., $\underline{56}$, 319-331.

Stephens, G.R. and Mill, D.E. 1973. In Proceedings Int. Conf. 'Land for waste management'. Ottawa, MRS. J. Tomlinson Ed., Canada, pp. 234-242.

THIRD SESSION

Chairman: T. Walsh

DISCUSSION

H. Laudelout *(Belgium)*

I would like to comment on nitrite accumulation because the problem of its transient presence is far from simple. Dr. van Veen mentioned yesterday that the transient accumulation of nitrite was not often observed but it will occur at high temperatures. Although ammonium inhibits nitrobacter activity, nitrite accumulation is not only due to ammonium inhibition, it is a combination of both temperature effect and ammonium inhibition. The conditions to observe this in soil are either a long incubation temperature or an additional reaction where nitrite ions are denitrified chemically with H-ions. So if you take, for instance, ammonium oxidation in acid tropical soils at a high temperature, nitrite accumulation will not be observed. The effect of temperature on the transient accumulation of nitrite has been studied experimentally (we have published two papers, one in French and one in English). It has a very practical interest for decreasing the nitrogen load of effluents because by the proper combination of a fairly high oxidation temperature and a high acidity, you can remove a large amount of nitrogen as gaseous product. The subject is fairly complicated and in modelling the possibility of nitrate/nitrite decomposition by H-ions should always be included.

E.A. Paul *(Canada)*

About two years ago in Uppsala at a SCOPE meeting, those present suggested that the agriculturalists were happy if they got rid of the nitrogen from the surface of their system. The meeting was worried about the N_2O in the atmosphere and many scientists were seriously worried about the possibility of the amount of nitrous oxide going into the atmosphere. I would like to hear the comments of the meeting as to whether there is a possibility of enough N_2O coming from some of the treatment plants going into the upper atmosphere and causing ozone depletion.

M.C. Masterson *(Ireland)*

I would like to make a general comment about the effect of temperature which was included in this morning's paper, and Professor Laudelout's remarks regarding the possible effects of different pathways of reactions at different temperatures. When we start to work out model systems, or when we want to do very precise work, we have to work in the laboratory, usually at laboratory temperatures, particularly those of us who live in temperate or cooler areas.

E.A. Paul

This brings up the important point of how easily can we adapt our data from one experiment to the other. For example, if Van Veen's model is valid you can run it at the laboratory temperature and extrapolate to the lower one. Eventually we have to be able to get to the state of working at one temperature knowing enough about what happens with change of temperature to enable us to predict. I think that temperature effects can now be predicted.

M.C. Cheverry *(France)*

In connection with Dr. Gasser's paper about the injection of slurry under grass, may I present my question to the meeting, perhaps to Dr. Starr. In modelling, how can we take into account the structural features of the soil?

J.L. Starr *(USA)*

I think that to fit this kind of thing into a model of the surface layer is very difficult, and perhaps it is not the right approach. If we have information on nitrogen uptake and the various nitrogen transformations that are averages for a given set of field conditions, then perhaps what we can do is to make various predictions. If we have some good estimates of the relevant soil parameters and relate these back to environmental influence - rainfall and irrigation - then I think we begin to develop a broader model that is useful, rather than one which tries to deal with specific problems.

G. Chisci *(Italy)*

I would like to respond to the question of spatial distribution in soil because it seems to me that any transformation must take into consideration the organic matter distribution in the soil.

T. Walsh *(Ireland)*

You have raised another point there that we must keep in mind. In the Drumlin soils in the north, which are intractable, difficult soils, in the summertime they will crack down 6 or 7 ft. If slurry is put on that soil it will enter the local drainage system very quickly. Therefore, an understanding of the actual structural conditions in some of these soils is necessary in order to understand how slurry application affects them.

A. Dam Kofoed *(Denmark)*

Agreed.

CONCLUSIONS AND RECOMMENDATIONS

In the absence of Mr. L'Hermite from the Commission, Dr. T. Walsh took the Chair. He opened the discussion by saying he wished the proceedings to be informal. He would be attending a meeting of SCAR the following day and would be able to carry the conclusions and recommendations of the meeting to that Committee. Mr. Dehandtschutter said that in Brussels they had already received a mandate from the SCAR to develop a new programme but it was necessary to have a meeting with the Expert Committee in order to establish this programme. Dr. Walsh then stressed that it would be helpful to the Expert Committee to have some guidance. Mr. Voorburg then commented that there are four members of the Expert Committee in this Workshop so that there would be direct communication.

In the ensuing discussion attention was paid to the need for good principles for land spreading and the development of models. This would require the ability to use the nutrients in the slurry effectively and avoiding pollution of water and smell problems. These aims raise the ideas of the optimum use of slurry and the protection of the environment. To meet both aims requires knowledge of what is happening when slurry is applied to land.

The problems of application of slurry in the field were discussed. How much can be applied safely? What are the effects of slurry on crop yield and composition? The economic factors need to be quantified, for example, how much will it cost to apply the slurry compared with the value of the nutrients in the slurry, which affects the attraction to the farmer of land spreading? We also need to consider the environmental implications of animal husbandry. How many animals can be kept within the given area in order to prevent pollution when the slurry from them is spread? A model could match the animal output to crop needs. This suggested that two types of models are required:

1. Systems models which allow the practical problems to be quantified.
2. Process models which describe the processes occurring when slurry is applied.

It was agreed that empirical models are better prepared locally. Models of wider interest on processes in soil should be of EEC interest.

There was general support for modelling.

It was agreed that the main model would be a suitable subject for a co-ordinated programme. The simpler models should be done immediately with discussion between interested workers.

Other points raised were the problems of the distribution of slurry and the land available for its disposal. The cost of transport is likely to be less than the cost of treatment if distances are less than 30 kilometres. If longer distances are involved the cost of transport may be greater than the value of the slurry. The effects of slurry on the structure of soil needs further investigation. This should be studied in the field as well as in the laboratory using available techniques for estimating the effects on soil micromorphology.

Several speakers stressed that slurry must go back onto the land. A hundred cows equal a thousand people for purposes of purification of slurry. It is necessary to relate the animal and crop production within the region. Models are required to allow use to be made of superfluous slurry within the region.

Other points
Simplified models are needed rather than simple models. This may be a difference of degree rather than of kind, and allows two approaches. The first is the empirical development of simple models and the second is the use of mathematical models to quantify and to simplify. The computer is of much

value in developing complex mathematical models.

The suggestion was made that the same experiments should be repeated in the field in different countries.

Dr. Walsh concluded the meeting by thanking Professor Laudelout for organising the seminar. Mr. Dehandtschutter wished to be associated with the thanks on behalf of Mr. L'Hermite and the EEC.

Conclusions

The meeting concluded that work on modelling should be continued with a high priority. They identified two types of programme required within this general framework.

1. There is a need for the production of simplified models of systems of slurry production and disposal, largely on a regional basis. This work would be financed regionally or nationally.

2. There is a great need for continuing work on the development of complex theoretical models dealing with the processes involved in slurry production disposal and particularly decomposition in soil.

These programmes have a world-wide application and within the EEC should form part of a co-ordinated programme.

LIST OF PARTICIPANTS

Mr. R. CALVET
Institut National de la Recherche Agronomique
Station de la Science du Sol
Route de Saint-Cyr
78000 Versailles
FRANCE

Dr. G. CATROUX
Laboratoire de Microbiologie des Sols
17 rue Sully, BV 1540
21034 Dijon Cedex
FRANCE

Mr. M. CHASSIN
Institut National de la Recherche Agronomique
Département Science du Sol
Route de Saint-Cyr
78000 Versailles
FRANCE

Mr. C. CHEVERRY
Ecole Nationale Supérieure Agronomique
de Rennes
Rue de Saint-Brieuc 65
35042 Rennes Cedex
FRANCE

Dr. G. CHISCI
Sezione di Fisica del Suolo
Istituto Sperimentale per la Studio e la
Fisica del Suolo
Piazza d'Azeguo 30
50121 Florence
ITALY

Mr. J. DEHANDTSCHUTTER
Commission of the European Communities
Division for the Co-ordination of
Agricultural Research
200 rue de la Loi
1040 Brussels
BELGIUM

Mr. C. DROEVEN
Station de Chimie et de Physique Agricoles
Chaussée de Wavre 115
5800 Gembloux
BELGIUM

Dr. J. DUFEY
Département Science du Sol
Université Catholique de Louvain
Place Croix du Sud 2
1348 Louvain-la-Neuve
BELGIUM

Dr. J.K.R. GASSER
Agricultural Research Council
160 Great Portland Street
London W1N 6DT
UK

Mr. J.C. GERMON — Laboratoire de Microbiologie des Sols
17 rue Sully, BV 1540
21034 Dijon Cedex
FRANCE

Mr. J.C. HAWKINS — National Institute of Agricultural Engineering
Wrest Park
Silsoe
Bedford MK45 4HS
UK

Mr. P. HERLIHY — The Agricultural Institute
Economic and Rural Welfare Centre
19 Sandymount Avenue
Dublin 4
IRELAND

Dr. R. KICKUTH — Gesamthochschule Kassel
Nordbahnhofstrasse 1a
3430 Witzenhausen
WEST GERMANY

Dr. A. DAM KOFOED — Agricultural Experimental Station
Askov
6600 Vejen
DENMARK

Mr. R. LAMBERT — Université Catholique de Louvain
Laboratoire de Physico-Chimie Biologique
Place Croix du Sud 2
1348 Louvain-la-Neuve
BELGIUM

Prof. H. LAUDELOUT — Université Catholique de Louvain
Département Science du Sol
Place Croix du Sud 2
1348 Louvain-la-Neuve
BELGIUM

Mr. R. LECOMTE — Centre de Recherches Agronomiques de Gembloux
Aveneu de la Faculté 22
5800 Gembloux
BELGIUM

Mr. C. MASTERSON — The Agricultural Institute
Soil Biology Department
Johnstown Castle
Wexford
IRELAND

Dr. E.A. PAUL — University of Saskatchewan
Department of Soil Science
Saskatoon, S7N 0W0
CANADA

Mr. A. RUELLAN	Ecole Nationale Supérieure Agronomique de Rennes Rue de Saint-Brieuc 65 35042 Rennes Cedex FRANCE
Mr. G. SOULAS	Laboratoire de Microbiologie des Sols 17 rue Sully, BV 1540 21034 Dijon Cedex FRANCE
Dr. J.L. STARR	The Connecticut Agricultural Experimental Station 123 Huntington Street Box 1106 New Haven Connecticut 06504 USA
Dr. J.A. VAN VEEN	Association Euratom - ITAL Keyenberg 6 Postbus 48 Wageningen THE NETHERLANDS
Prof. Dr. H. VETTER	Landwirtschaftliche Untersuchungs- und Forschungsstelle der Landwirtschaftskammer Weser - Ems 2900 Oldenburg WEST GERMANY
Mr. J.H. VOORBURG	Rijks Agrarische Afvalwaterdienst Kemperbergerweg 67 Arnhem THE NETHERLANDS
Dr. T. WALSH	The Agricultural Institute 19 Sandymount Avenue Dublin 4 IRELAND

Recording service:

Mrs. MOLLY ROBINS	Janssen Services 14 The Quay Lower Thames Street London EC3R 6BU UK

MANUSCRIPT PREPARED BY JANSSEN SERVICES, LONDON.

DATE DUE